T0255239

Machine Learning For Network Traffic and Video Quality Analysis

Develop and Deploy Applications Using JavaScript and Node.js

Tulsi Pawan Fowdur
Lavesh Babooram

Apress®

Machine Learning For Network Traffic and Video Quality Analysis: Develop and Deploy Applications Using JavaScript and Node.js

Tulsi Pawan Fowdur
Department of Electrical and Electronic
Engineering, University of Mauritius,
Reduit, Mauritius

Lavesh Babooram
Reduit, Mauritius

ISBN-13 (pbk): 979-8-8688-0353-6
https://doi.org/10.1007/979-8-8688-0354-3

ISBN-13 (electronic): 979-8-8688-0354-3

Copyright © 2024 by Tulsi Pawan Fowdur, Lavesh Babooram

This work is subject to copyright. All rights are reserved by the publisher, whether the whole or part of the material is concerned, specifically the rights of translation, reprinting, reuse of illustrations, recitation, broadcasting, reproduction on microfilms or in any other physical way, and transmission or information storage and retrieval, electronic adaptation, computer software, or by similar or dissimilar methodology now known or hereafter developed.

Trademarked names, logos, and images may appear in this book. Rather than use a trademark symbol with every occurrence of a trademarked name, logo, or image we use the names, logos, and images only in an editorial fashion and to the benefit of the trademark owner, with no intention of infringement of the trademark.

The use in this publication of trade names, trademarks, service marks, and similar terms, even if they are not identified as such, is not to be taken as an expression of opinion as to whether or not they are subject to proprietary rights.

While the advice and information in this book are believed to be true and accurate at the date of publication, neither the authors nor the editors nor the publisher can accept any legal responsibility for any errors or omissions that may be made. The publisher makes no warranty, express or implied, with respect to the material contained herein.

Managing Director, Apress Media LLC: Welmoed Spahr
Acquisitions Editor: Celestin Suresh John
Development Editor: Laura Berendson
Coordinating Editor: Gryffin Winkler
Copy Editor: April Rondeau

Cover designed by eStudioCalamar

Cover image by Joshua Fuller on Unsplash (www.unsplash.com)

Distributed to the book trade worldwide by Apress Media, LLC, 1 New York Plaza, New York, NY 10004, U.S.A. Phone 1-800-SPRINGER, fax (201) 348-4505, email orders-ny@springer-sbm.com, or visit www. springeronline.com. Apress Media, LLC, is a California LLC, and the sole member (owner) is Springer Science + Business Media Finance Inc (SSBM Finance Inc). SSBM Finance Inc is a **Delaware** corporation.

For information on translations, please e-mail booktranslations@springernature.com; for reprint, paperback, or audio rights, please e-mail bookpermissions@springernature.com.

Apress titles may be purchased in bulk for academic, corporate, or promotional use. eBook versions and licenses are also available for most titles. For more information, reference our Print and eBook Bulk Sales web page at http://www.apress.com/bulk-sales.

Any source code or other supplementary material referenced by the author in this book is available to readers on GitHub (https://github.com/Apress). For more detailed information, please visit https://www.apress.com/gp/services/source-code.

If disposing of this product, please recycle the paper

Table of Contents

About the Authors

Dr. Tulsi Pawan Fowdur received his bachelor of engineering degree in electronic and communication engineering with honors from the University of Mauritius in 2004. He was also the recipient of a gold medal for having produced the best degree project at the Faculty of Engineering in 2004. In 2005, he obtained a full-time PhD scholarship from the Tertiary Education Commission of Mauritius and was awarded his PhD in electrical and electronic engineering in 2010 by the University of Mauritius. He is also a registered chartered engineer of the Engineering Council of the United Kingdom, fellow of the Institute of Telecommunications Professionals of the United Kingdom, and a senior member of the IEEE. He joined the University of Mauritius as an academic in June 2009 and is presently an Associate Professor at the Department of Electrical and Electronic Engineering of the University of Mauritius. His research interests include mobile and wireless communications, multimedia communications, networking and security, telecommunications applications development, the Internet of Things, and artificial intelligence (AI). He has published several papers in these areas and is actively involved in research supervision, reviewing papers, and also organizing international conferences.

Lavesh Babooram received his bachelor of engineering degree in telecommunications engineering with networking with honors from the University of Mauritius in 2021. He was also awarded a gold medal for having produced the best degree project at the Faculty of Engineering in 2021. Since 2022, he has been pursuing a master of science degree in applied research at the University of Mauritius. With in-depth knowledge of telecommunications applications design, analytics, and network infrastructure, he aims to pursue research in networking, multimedia communications, Internet of Things, artificial intelligence, and mobile and wireless communications. He joined Mauritius Telecom in 2022 and is currently working in the Customer Experience and Service Department as a pre-registration trainee engineer.

About the Technical Reviewer

Kamalakshi Dayal received a bachelor of engineering degree in telecommunications engineering with networking, with first-class honors, from the University of Mauritius in 2021. She then undertook a one-year internship at Huawei Technologies (Mauritius) Ltd., whereby she had the opportunity to work on two highly innovative and ground-breaking award-winning projects, assessed at the regional level. She is currently an engineer at the company and is specializing in a multitude of product lines, mainly in convergent billing systems, IPTV, and cloud services.

CHAPTER 1

Introduction

This chapter introduces the concepts of network traffic monitoring and analysis (NTMA) and video quality assessment (VQA). It discusses the significance of NTMA and VQA in modern telecommunications by emphasizing the need to achieve optimal network performance and to boost user experience, which encompasses both the flow of network parameters and the video streaming quality. The sections in this chapter set the tone for the implementation phase by first exploring the need for NTMA and VQA with regard to preserving the quality of service (QoS) in different network environments. With the aim of executing these processes on any device irrespective of the operating system, a combination of machine learning (ML), JavaScript, and Java, with Node.js and Apache HTTP Server as the backend frameworks, is used. Network traffic parameters are read from the local device before streamlining client–server interactions to produce meaningful prediction and classification results. Likewise, a combination of distortion artifacts is applied to a playing video to produce a video quality score. The following sections provide an overview of the book's structure and offer a glimpse into what will be covered in the next chapters.

1.1 Overview of Network Traffic Monitoring and Analysis

NTMA is a cornerstone of today's telecommunications environment, essential for ensuring network efficiency, security, and user experience. The exponential rise of digital interactions, exacerbated by the introduction of 5G and the Internet of Things (IoT), emphasizes the need to monitor and regulate network traffic patterns. NTMA serves as a watchdog in the networking biosphere, actively monitoring the changing patterns of data streams. This heightened surveillance is motivated by the primary goals of assessing traffic conditions, identifying anomalies, and improving performance to provide consumers with a smooth digital experience [1].

© Tulsi Pawan Fowdur, Lavesh Babooram 2024
T. P. Fowdur, L. Babooram, *Machine Learning For Network Traffic and Video Quality Analysis*,
https://doi.org/10.1007/979-8-8688-0354-3_1

Statistics from industry publications shine a bright light on the scope of NTMA. According to Cisco's Annual Internet Report, worldwide IP traffic was expected to triple to 396 exabytes per month by 2022, a threefold increase from 2017 [2]. In an era when a massive amount of multimedia content is consumed at lightning speed, NTMA proves to be a fundamental aide for telecommunications operators as they not only decode intricate traffic trends but also anticipate and mitigate network congestion, bottlenecks, and possible security breaches.

Within the NTMA paradigm lies a whole range of complexities, starting with the diverse nature of internet traffic. This jumbled and disorganized mix contains packet headers, payload data, and metadata, all originating from different multimedia streams. Putting this mosaic of information under the analytics microscope offers insights into the transmission dynamics among devices, applications, and services, shedding light on patterns that might otherwise go unnoticed and unreported. However, it is computationally and mathematically intensive to yield meaningful information from the colossal amount of traffic that traverses a typical network, often exceeding billions of packets per second. Adding to the list of hurdles is the encrypted nature of communication packets, which hides critical information, thus further necessitating complex approaches for useful analysis.

The real-world ramifications of NTMA are numerous and multifaceted. For example, NTMA serves as a pillar in network operations centers (NOCs) and service operations centers (SOCs), where it assists in the rapid identification of harmful activity, the prevention of data leaks, and the protection of sensitive assets [3]. Likewise, NTMA improves quality of service (QoS) concerning network optimization by dynamically and proactively monitoring and reporting network congestion sites and rerouting traffic. Another poll by Spiceworks indicated that 53 percent of information technology (IT) professionals perceive NTMA to be the most pivotal component of maintaining their infrastructure, highlighting the need for service providers to evaluate user experiences in real time and guarantee seamless connectivity [4].

As NTMA transitions through this period of heightened technological change, machine learning (ML) has emerged as a powerful ally. ML propels NTMA to previously unattained heights, using cutting-edge algorithms to fuel predictive abilities, robust anomaly detection, and responsive network management. On the same wavelength, the adaptability of ML frameworks with regard to being lightweight, compact, fast, and efficient acts as the pillar upholding the capability of the typical end user to perform

NTMA at the client's side. This revolutionizes digital ecosystems, allowing the end user to be equipped with high-performing algorithms to gauge the reception quality of multimedia content, resulting in a more sophisticated and complete ecosystem.

NTMA can be represented as a layer between network components and SOC and NOC platforms, as illustrated in Figure 1-1.

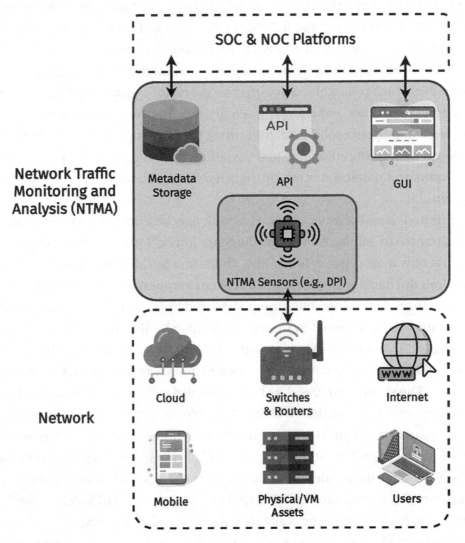

Figure 1-1. *NTMA layer in modern frameworks*

The following sections elaborate upon the different aspects that make up NTMA, such as its necessity and perks.

1.1.1 Importance of NTMA

NTMA stands out like a sentinel in the complex web of today's telecommunications, keeping tabs on the virtual veins that supply the interconnected world. As digital interactions become ubiquitous, the sheer quantity of data moving across networks has skyrocketed. International Data Corporation (IDC) estimates that worldwide internet users will create 175 zettabytes by 2025 [5]. NTMA is the guardian of network security, privacy, and efficiency in the face of an ever-increasing data flood.

One of the primary goals of NTMA is to make sense of the random nature of network traffic, which from a global perspective is a chaotic jumble of different kinds of interactions, data flows, and endpoints. Firstly, by describing these patterns, NTMA offers network managers a solid base from which to begin improving performance, and secondly, by spotting anomalies and deviations from the preset norms, NTMA ensures alarms and notifications can be triggered during possible security breaches and interruptions [6].

A distributed denial of service (DDoS) assault may be imminent if data packets suddenly increase in volume from a specific geographical location. The strength of NTMA is not only in its capacity to spot this aberration but also in its readiness to act swiftly to limit the damage it causes. The threat environment is changing, with the number of recorded DDoS assaults having increased by 67 percent in 2022 compared to the previous year, as reported by Cloudflare. Kaspersky also registered nearly 58,000 such incidents [7]. As such, NTMA serves as a digital lighthouse, allowing network experts to remain vigilant against new threats and quickly restore normality after a compromise. The sheer volume of data being sent also calls for strategic use of resources and careful network optimization. With NTMA's help, ISPs and other telcos may pinpoint impediments, ease traffic, and distribute bandwidth more effectively. This is significant because the demand for telecommuting jobs, online courses, and digital media has been on the rise ever since the pandemic in 2020. As a result of NTMA's work to improve QoS, customers can enjoy a more consistent digital experience, with HD video streaming, page loads, and app responsiveness all functioning at peak efficiency.

To sum up, NTMA is more than just a monitoring process; it also improves network security and makes users satisfied. Its pervasiveness is highlighted by the prevalence of digital interactions, and its significance will grow as networks mature. NTMA paves the way for a digitally linked society that lives on efficiency and dependability by welcoming the diversity of network traffic, comprehending its subtleties, and converting them into useful information.

1.1.2 Key Objectives of NTMA

At its core, NTMA aims to tick several boxes of network management and optimization. As the number of digital contacts continues to grow, with a recorded 5.44 billion internet users as of April 2024, according to Statista [8], NTMA will play an increasingly important role in ensuring the continued viability, safety, and efficiency of today's networks. Strategically, NTMA aims to do the following [9]:

- Describe typical traffic flows and data streams at the application level.

- Identify outliers and anomalies proactively.

- Enhance overall efficiency through optimal routing.

- Observe traffic trends and forecast traffic for different network environments.

- Provide a bird's-eye view of the network.

Telecommunications experts may get invaluable insights into the patterns that characterize network activity thanks to NTMA's rigorous categorization and analysis of the various data streams crossing networks. Anomaly detection, the pre-emptive detection of abnormalities from established traffic standards, relies on this familiarity with normality as its foundation. For instance, VMware Carbon Black reported a 118 percent rise in assaults against banking firms in 2020 [10]. The ability of NTMA to detect abnormalities, including unexpected increases in traffic or changes in communication patterns, is crucial to preventing security breaches. As the quantity and significance of digital interactions grow, NTMA is a natural fit with the need to optimize networks. To improve QoS and resource management, NTMA identifies areas of overcrowding, traffic limitations, and other performance constraints. This is especially important now since the average internet speed throughout the world is expected to reach 110.4 Mbps by 2023, as reported by Ookla [11]. Users now demand consistent, low-latency, high-speed access, which is upheld by NTMA's features. NTMA is thus the nexus at which security, efficiency, and user experience all meet, and addresses the complexities of network dynamics. The goals can be summarized as protecting the delicate balance between the ever-increasing amount of data and the need for a seamless user experience, which reverberates more strongly as digital ecosystems evolve.

1.1.3 Network Traffic Components

The backbone of today's communication infrastructure, network traffic is itself a complex phenomenon. Internet Protocol (IP) packet headers offer a background for data transfer by encoding information such as the source and destination addresses and port numbers. Metadata, which provides background information, supplements payload data, which represents the message itself. The wide variety of streaming services, signaling packets, and control messages adds to the already chaotic state of network traffic [12]. The IDC predicts that by 2025 the global datasphere will grow to 175 zettabytes, 75 percent of which will relate to non-PC devices, in turn reflecting the growth of non-PC traffic sources [13]. By dissecting these complex parts, network administrators may learn more about user behaviors, application relationships, and traffic flows. Underneath the surface of frictionless digital exchanges is a tangled web of network traffic. This complex network consists of a symphony of parts, all of which play a role in the seamless transfer of information between computers. Statista predicts that there will be 29.4 billion IoT devices in use throughout the globe by 2030, which means that there will be an exponentially increasing variety and number of network traffic components [14].

Data packets are the building blocks of network traffic. These digital packages transmit data to their final destinations. Like letters in the mail, these packets have unique "headers" that include crucial information, such as where they came from and where they're going, as well as the ports and protocol that they are using. Routers and switches may use this contextual information to send data to the correct destinations. The "payload," or body of a message, is contained even deeper inside these packets. The information itself, such as text, photos, video streams, or application-specific data, is located here during transmission. The foundation of today's digital experiences rests on the fast-paced communication of these packets. Figure 1-2 shows the structure of an IP packet [15].

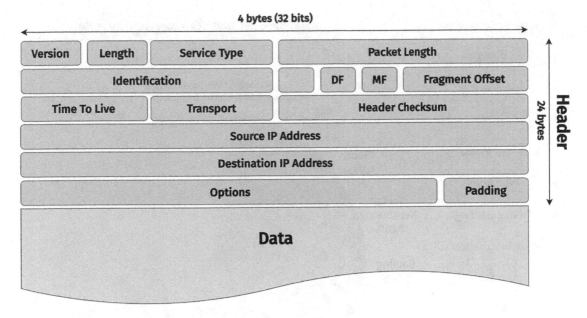

Figure 1-2. *Structure of an IP packet*

Likewise, "flows" of connected packets that share characteristics are highlighted in the network [16]. Network managers can thus comprehend communication dynamics between apps, services, and devices—i.e., at different levels—leading to a bird's-eye view of the network dynamics. This is where the multiverse of stream types also comes into play, where Voice over Internet Protocol (VoIP) conversations, video streaming, and online surfing activities can be segmented and analyzed according to their unique trends. In addition, metadata of packet timestamps, communication length, and transmission sequence numbers further, among others, help to enhance and grasp network interactions.

However, the wide spectrum of network traffic elements poses certain difficulties. The expansion of video media in internet traffic, which Cisco predicted, would account for 82 percent of all IP traffic by 2021 [17] and presents complexities that necessitate advanced traffic analysis methodologies. The complex dance of packet headers, payload data, metadata, and multimedia feeds produces an evolving complex terrain that NTMA attempts to maneuver. Understanding these components is essential for gaining comprehension of user habits, recognizing causes of network congestion, and guaranteeing an unparalleled user experience in a progressively data-driven environment. As of January 2023, Sandvine reported a 24 percent increase in video traffic, with Netflix overtaking YouTube in terms of video consumption. A broad list of traffic categories, together with the total volume by traffic category consumed worldwide by the end of January 2023, is depicted in Figure 1-3 [18, 19].

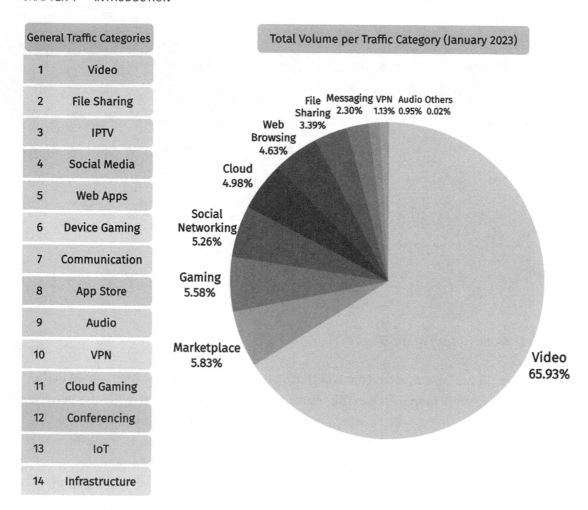

Figure 1-3. *General network traffic categories and total worldwide volume consumption per category*

1.1.4 NTMA Techniques and Methodologies

NTMA serves as proof that technological innovation and analytical prowess may successfully coexist. From simple packet capture to powerful ML algorithms, NTMA marches through the raw data maze by allowing telecom engineers and data analysts to mine for useful patterns, irregularities, and conclusions. The concept of capturing packets is at the pinnacle of NTMA techniques. This fundamental approach intercepts and captures data packets as they travel through a network. Network experts obtain open-to-use access to the information contained in packets by capturing them in their natural form, from headers through payloads [20]. Packet capture is the foundation for understanding communication patterns, analyzing data flow features, and, most important, recognizing abnormalities that may indicate safety issues or connection inconsistencies. This method is analogous to a digitized Rosetta Stone, deciphering the meaning of network connections.

One of the core methods of network analysis revolves around the handling of large floods of network data through a flow-based evaluation, where flows are formed by grouping packets according to shared features, such as source and destination addresses, port numbers, and protocol types [21]. This decoupling allows network managers to make better choices based on accumulated information rather than on individual packets, resulting in a more holistic knowledge of network behavior [22]. The insights are then developed and derived from the behavior of patterns observed at the device, application, and service levels.

As NTMA professionals explore this terrain, two different methods emerge: active network monitoring and passive network monitoring. To measure throughput, safety, and dependability, active network monitoring generates traffic on purpose. Active monitoring is a method of evaluating the state of a network by simulating different situations via the use of "probes" or "agents" to collect data on response times, latency, and other metrics [23]. Similar to a lab experiment, this method allows for unobstructed views into the structure and operation of a network. When availability and responsiveness are mission-critical, active monitoring shines because it allows for a continuous evaluation of the network's behavior and performance. MarketsandMarkets predicts that the worldwide network monitoring market will surpass $6.97 billion by 2030 [24], and the financial industry, for one, uses active monitoring to guarantee that trading platforms react quickly to fluid market developments.

Passive network monitoring, on the contrary, takes the role of a quiet observer, documenting all network activity without adding to the existing volume of data being sent [25]. Techniques like deep packet inspection (DPI) exemplify this approach by letting network managers record and examine traffic in its unaltered condition. When deep historical context into a network's behavior is needed, passive monitoring is preferred. To better allocate resources and optimize their networks, internet service providers (ISPs) often use passive monitoring techniques, such as analyzing patterns in traffic [26].

Moreover, the extra layer of reliability, and high accuracy, is fueled by the adoption of data analysis and ML, which elevate NTMA frameworks to a whole new dimension. The array of ML algorithms provides a comprehensive toolset for recognizing patterns and anomaly detection [27]. Network experts can detect abnormalities that would otherwise go undetected if inference and descriptive statistics were not used [28]. Additionally, predictive analytics and anomaly detection have entered a new age due to the partnership of NTMA and ML. From supervised algorithms to deep learning (DL) architectures, ML models consume enormous datasets to unearth previously hidden correlations and outliers. Having ascended to reach a staggering $6 trillion loss in global cybercrime in 2021, cyberattacks are expected to rise further [29]. This integration with ML can improve threat detection accuracy. Since ML models can change and adapt from erratic network behaviors, NTMA is better able to foresee potential problems, spot outliers, and suggest preventive measures. ML transforms NTMA from a reactive approach to a proactive one, meeting the needs of a modern, data-driven society.

As NTMA advances into an age defined by an overload of information and advanced technology, its methods reflect this rapid transformation. The complex dynamics of network traffic are revealed through the orchestration of insights gained through packet capture, flow-based analysis, statistics, and ML. In an era characterized by connection and knowledge, this symphony arms telecommunications experts with the means to not only understand but also predict network behavior, guiding networking infrastructures toward efficiency, security, and performance. Figure 1-4 depicts the conceptual framework that serves as the foundation for the majority of recent studies on NTMA proceedings [30].

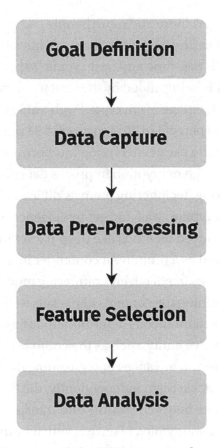

Figure 1-4. *General framework for NTMA procedure*

1.1.5 Challenges of NTMA

NTMA is a sentinel amidst the ocean of data flowing through the complex web of today's networks, entrusted with extracting the hidden storylines lying within the massive influx. However, this area of knowledge conceals a maze of obstacles that can only be overcome with creativity, flexibility, and cutting-edge approaches.

One of the primary difficulties is the overwhelming quantity of network traffic [31]. Data centers globally handle an estimated 100 zettabytes of content yearly, and in this day and age, an individual user creates terabytes of data yearly, meaning that the amount of information crossing networks is astonishing [32]. To make sense of this deluge of information, sophisticated methods, such as data reduction, aggregation, and selective sampling, are required. The likelihood of missing valuable insights also increases as more data becomes available. As more and more people use streaming services, cloud technology, and IoT devices, the diversity of network interactions increases, necessitating the analysis of both real-time and past events [33].

A new level of complication is being added by the shift to encrypted transmission. Google has reported that the percentage of encrypted web traffic on the internet increased from approximately 50 percent in 2014 to around 95 percent after 2020 [34], demonstrating how widespread the use of encryption has become in response to growing security concerns. Although encryption improves data privacy and security, it also hides packet contents, making deciphering them as difficult as deciphering coded letters. To break through the encryption wall and get intelligence without breaching data protection regulations, advanced methods like deep packet inspection (DPI) are required. DPI is widely recognized as a prominent contender in the field of web security, exhibiting considerable capabilities in effectively countering modern web-based threats. It is a process that involves examining the contents of packets as they traverse a specific location and afterward making instantaneous determinations based on predefined criteria established by the respective entity, such as the company, ISP, or network administrator [35]. Despite the recent advances in segregating the building blocks of network traffic in an attempt to tackle them independently, the NTMA community still has great trouble striking a balance between security and transparency, especially with the usage of decryption and user anonymity. Likewise, with an expected 41.6 billion devices—including mobile phones, tablets and IoT sensors—expected by 2025 as forecasted by IDC [36], it is becoming more difficult to standardize analytic procedures across devices and communication protocols. This necessitates the development of methods that can handle the nuances of multimodal traffic.

1.1.6 Use Cases of NTMA

The protection of vital digital assets is NTMA's top priority as the threat environment shifts and assaults become more sophisticated. To detect security breaches or malicious actions, NTMA analyzes network traffic patterns for deviations. The company Cybersecurity Ventures estimates that yearly expenditures related to cybercrime will amount to $10.5 trillion by 2025, making constant monitoring a crucial line of defense [37]. Network security experts may use NTMA's findings to strengthen protections, identify breaches, and react swiftly to new threats like distributed denial of service (DDoS) assaults and ransomware penetration [38].

When it comes to network optimization, which is crucial in today's digital economy where constant connection and fast data transfer are paramount, NTMA examines network traffic patterns to identify hotspots for congestion, as well as limitations in bandwidth and performance [39]. From this vantage point, IT managers can more effectively distribute network assets, enhance users' quality of service (QoS), and prevent service disruptions. With Ericsson predicting that 5G networks will reach 45 percent of the world's population by 2024, it is clear that effective network management is essential to satisfy consumer expectations as the market for fast and low-latency internet connectivity continues to rise [40].

Users may expect a flawless online experience thanks to NTMA's effective supervision of performance bottlenecks and optimization of network resources. This is evident in customer behavior where those who encounter website performance difficulties are less inclined to return to the site. Users' digital engagements are significantly impacted by NTMA's role in ensuring network flexibility and effectiveness, whether they are downloading high-definition material, holding virtual meetings, or making real-time online transactions [41].

More of NTMA's use cases are highlighted as follows:

- Network traffic characterization and monitoring

- Network optimization and planning

- Detection of network security violations

- Evaluation and improvement of network QoS

- Predictive maintenance of network components

- Forensic analysis

- Compliance and regulatory reporting

- Vulnerability scanning and patch management

- Cost optimization

1.1.7 Emerging Trends in NTMA

The incorporation of AI and ML into NTMA has been identified as a game-changing development. Artificial intelligence–driven insights offer important foresight about network activity in light of the daily data deluge. Companies like Cloudflare use AI to analyze massive volumes of network traffic data to identify and mitigate distributed denial of service (DDoS) attacks in real time. With the ever-increasing computational power at their disposal, NTMA is now able to not only spot patterns and abnormalities, but also predict them, in turn greatly improving network security and performance.

Edge computing's rise is also influencing changes to NTMA's methodology. NTMA has evolved from centralized evaluation to real-time monitoring as more and more IoT devices collect data at the network's edge. Pioneers of edge computing like Siemens use it in production, where material gathered from sensors implanted in equipment is evaluated locally for instantaneous decision-making [42]. This development allows NTMA to quickly detect problems and enhance performance as data is created, which helps to lessen delays and guarantee smooth user experiences.

In addition, the growing popularity of cloud computing is dramatically improving NTMA's versatility and accessibility [43]. Network operators now have the flexibility to install and administer remote monitoring throughout scattered environments with the help of cloud-based NTMA services like those provided by ThousandEyes, which is now part of Cisco [44]. This adaptability serves the ever-changing nature of network topologies, including the popularity of telecommuting. The move to NTMA in the cloud is an example of how the modularity and remote availability of the cloud meet the ever-changing needs of today's networks. These developments work together to provide NTMA with new capabilities, allowing it to flourish in the age of big data and digital complexity. NTMA plots a trajectory toward dynamic, smart, and proactive network management by making use of AI and ML, adopting edge computing, and integrating cloud solutions. Changes like this highlight NTMA's lasting impact on the reliable, high-performance networks that power the modern digital world [45].

1.1.8 Bridging the Gap between NTMA and User Experience

NTMA has emerged as the foundation of network responsiveness, durability, and ultimately user experience as the digital age has progressed. NTMA and the quality of experience (QoE) of the end user are intrinsically linked [46]. The capacity of NTMA to analyze network traffic, pinpoint processing delays, and enhance QoS immediately translates into improved customer experience. Consider the case of a streaming platform that is monitored in real time with NTMA. Service monitoring and management teams can keep tabs on the status of the network and act quickly if there is a drop in streaming quality, which is often summarized as a mean opinion score (MOS) obtained through video quality assessment (VQA). This instantaneous adaptation goes hand in hand with viewers' evolving habits. For example, Netflix reported an increase of 15.77 million new subscribers in the first quarter of 2020, an increase of 23 percent over the same period in 2019 [47]. NTMA aids in client satisfaction and retention by guaranteeing a problem-free streaming experience.

In addition, NTMA's expertise applies to e-commerce, where customers' interactions have a major impact on their final purchases. The effectiveness of the network is crucial in the current environment. Potential latency concerns are identified by NTMA's study of user interactions and transaction flows, guaranteeing lightning-fast, error-free exchanges. Given that Statista predicts that global e-commerce sales will hit $5.56 trillion by 2027 [48], this is of utmost importance. Beyond its role in technical management, NTMA also helps drive revenue by improving the quality of the user experience.

Additionally, remote work scenarios benefit from the integration of NTMA and user experience. The rising popularity of working from home has made network speed and reliability more important than ever. Because of NTMA's efforts to reduce lag time, improve the quality of video conferences, and guarantee the timely transmission of data, dispersed teams can work together effectively. An overwhelming majority of remote workers, amounting to 58 percent, agree that telecommuting will increase in popularity over the next decade, as per research conducted by Owl Labs [49]. Due to the growing popularity of telecommuting, NTMA must prioritize QoE of streaming and video conference applications. To put it simply, NTMA bridges the gap between the technology economy and human experience. Improved user experience, engagement, and productivity are all direct products of NTMA's ability to analyze network dynamics and uphold performance.

1.2 Overview of Video Quality Assessment

The quality of video material has become a major differentiator in the complex digital fabric of contemporary multimedia consumption. Beyond the domain of basic technical standards, video quality assessment (VQA) stands as a key field that tries to objectively quantify and subjectively comprehend the perceptual quality of video output. VQA is, at its core, an interdisciplinary field that draws from engineering, psychology, and human-computer interaction. The ultimate objective is to level the playing field between the technical aspects of video, such as resolution, bit rate, and compression artifacts, and the subjectivity of human viewers. This effort recognizes the intricacies of the human visual system, which include the interaction of cognitive processes, visual acuity, and psychological aspects that all contribute to the unique ways in which people take in and make sense of the world around them [50]. The rapid spread of streaming video across several channels highlights the need for VQA. It is crucial that the videos people watch online not only work technically but also captivate and immerse them. In this sense, VQA is the map that helps content producers, streaming services, and telecoms meet consumers where they want them to be in terms of the quality of the experiences they get.

Changes in the digital world have led to corresponding shifts in the variables that determine video quality. While extremely important, video quality factors like resolution and bit rate are only a small piece of the issue. Many elements can affect how an audience perceives a video, including compression artifacts, color correctness, motion fluidity, and audio synchronization. Thus, VQA incorporates both objective and subjective evaluation approaches, with the goals of measuring technical features and recording viewers' emotional responses.

The dynamic development of VQA may be seen in the shift from elementary, rule-based algorithms to sophisticated ML models. ML approaches may now be used to replicate human perceptual judgments, replacing older methods based on measures such as peak signal-to-noise ratio (PSNR) and structural similarity index (SSIM). In recent years, deep learning architectures such as convolutional neural networks (CNNs) and recurrent neural networks (RNNs) have become extremely useful tools for gleaning subtle but important visual details from video [51].

In addition to its use in the entertainment industry, VQA has practical implications in the medical, security, and academic sectors. In medical imaging, for instance, an incorrect representation of diagnostic images might have fatal consequences. VQA reduces the potential for misunderstanding by making sure that medical images are true to their origins. With VQA, security professionals can be certain that they are viewing authentic, unedited video feeds, which assists in the discovery of abnormalities. VQA's continued exploration of AI, ML, and the psychology of perception has made it a centerpiece of the modern digital experience. It solidifies its role as a medium for providing not only videos, but also immersive visual tales that capture, connect, and reverberate with users by aligning technical requirements with human perception.

1.2.1 Significance of VQA

With video content sitting at the throne of current multimedia consumption trends, VQA is the heart that brings together content development and distribution strategies by molding user experiences through optimized metrics. VQA aims to answer questions such as how much of an impact video content quality has on audience engagement and satisfaction. Content-making industries such as internet protocol television (IPTV) broadcasters and over-the-top (OTT) streaming platforms require robust monitoring programs that display the current status of media being transmitted with regard to availability, running time, service quality, and user experience, which all form part of their feedback loop. This industry understands that viewers don't simply consume content; they live it. This section looks into the critical importance of VQA in today's multimedia ecosystem.

The direct effect of VQA on user engagement is likely the clearest indicator of its importance. The quality of the video content is the first line of interaction for digital customers in an age of abundant choices and short attention spans. This highly resonates with the thumping rise of short and captivating video consumption, among examples such as TikTok, YouTube Shorts, and Instagram Reels [52]. VQA operates as a watchdog, checking videos for flaws that might cause the viewer to be taken out of the experience, such as distortions, glitches, and other deficiencies. VQA ensures that videos have no distracting flaws, setting the stage for instant audience engagement. Among the advantages of VQA are its monetary effects. According to the *Harvard Business Review*, a 25 percent to 95 percent boost in profitability may be achieved by retaining just 5 percent more customers, leading to a direct correlation between content quality and viewer happiness [53].

The relevance of VQA goes beyond technical prowess to the psychological resonance of video. Videos are a powerful medium for sharing ideas, feelings, and experiences. VQA works to maintain the emotional resonance that gives videos their power, whether they are uplifting commercials, informative tutorials, or riveting narratives. Users have increasing expectations for video quality due to the proliferation of high-definition monitors, augmented reality, and virtual reality. User-created videos and live streaming have helped to democratize content creation, but this has led to a corresponding need to democratize quality. VQA's duty also includes analyzing content from many publishers to guarantee a constant, high-quality visual experience independent of the video's original source. Its relevance is far reaching and affects not just user engagement but also content developers and producers. VQA protects interest, financial gain, and emotional impact in the digital era by promoting high standards for video. As a result, the process of making and consuming digital stories is elevated to the level of an artistic symphony between technology and human perception, further enhancing the digital mosaic that characterizes contemporary encounters.

1.2.2 Factors Affecting Video Quality

The experience of watching a video is influenced by a wide variety of circumstances, each of which plays a unique function in the overall impression that the audience forms. From technical parameters to perceptual psychology, researchers break down what makes for a satisfying video-viewing experience, as shown in Table 1-1 [54].

Table 1-1. *Factors Affecting Video Quality*

Factor	Description
Resolution	It stands for the number of pixels in each dimension and is the primary determinant of video quality. Clearer and more detailed images may be seen at higher resolutions like 4K and 8K. The resolution of a video has a major effect on how it appears on a screen, especially a very big one. However, greater bandwidth and computing power are required for higher resolutions.
Bit Rate	The pace at which bits are exchanged in a certain length of time. As the bit rate increases, more visual details may be sent with less need for compression, thus impacting video quality. High bit rates can put a load on a network, so finding the right middle ground is tricky.
Compression	Techniques for compressing video, such as H.264 and H.265, are crucial for saving space and bandwidth while storing and transmitting video. However, blockiness or blurring can be introduced by overly strong compression, impairing video quality.
Frame Rate	The number of frames exhibited each second, also known as "fps," is governed by the frame rate. Frame rates of 24, 30, and 60 per second are the norm. The fluidity of action is improved by frame rates of 120 or 240 fps and higher. However, different audiences have different preferences when it comes to frame rate.
Color Accuracy	Realistic visual experiences rely heavily on color reproduction. Colors that aren't accurate might make images look fake or unpleasant. In fields such as medical imaging and filmmaking, color fidelity is essential.

(*continued*)

Table 1-1. (*continued*)

Factor	Description
Motion Handling	Fast motion rendering without distortion is a must for any scenario with a lot of activity or for any sports broadcast. The ability of displays to process motion is tied to their frame rate, refresh rate, and reaction time specifications.
Audio Quality	Video footage is enhanced with high-quality audio. Audio encoding, stereo/surround sound systems, and noise cancellation methods all contribute to overall sound quality. The audio and visual elements of a presentation must be in sync for optimum viewing experience.
Content Source	Quality control measures are in place for most commercially created content, while user-generated content can be all over the place. A video's expected quality thus depends largely on its source.
Viewing Environment	Perceived video quality is also affected by the viewer's surroundings. This includes lighting conditions, screen size, viewing distance, and background noise. Glare from windows, for instance, might harm the overall impression.
Psychological Factors	Audience expectations, focus, and mood all have a part in how humans interpret a video's quality psychologically. However, technical defects in highly anticipated content may be more visible if they aren't compensated for by compelling content.
Device Capability	The viewing experience is dictated by the device being used to watch a video, ranging from a hand-held device such as a smartphone to a high-end television with 4K capabilities.

The previously mentioned parameters thus affect the final video output. Some possible effects are as follows [55]:

- **Blocking**: The process of video coding involves the use of block-based techniques. Consequently, degradations in network performance may lead to loss of data or coding errors.

- **Blurring**: The phenomenon of blurring is evident when there is a reduction in spatial information or distinct features, resulting in the blending and indistinctness of boundaries surrounding a scene or object.

- **Edginess:** This pertains specifically to the differences observed in the edges when comparing the modified video to its original version. The objects present in the content exhibit uneven edges.

- **Motion Stutter:** The occurrence of motion stutter is typically assessed by comparing real-time measurements to video time using sequence numbering. In some cases, content freezing or segment skipping can occur.

Content creators, as well as service providers, must understand and assess the impact of these factors in order to provide a satisfactory, immersive, and striking experience for humans.

1.2.3 Evolution of VQA Approaches

Before the development of advanced ML and DL techniques, VQA was based on traditional methods. In this section, the evolution of VQA approaches is elaborated [56], as follows:

- **Subjective Assessment:** The backbone of classic VQA, i.e., subjective evaluation, relies on human observers to provide ratings to videos based on their own experiences. The Absolute Category Rating (ACR) approach is an example of a subjective test that uses human judgment to evaluate video quality. Ratings and labels given by viewers reveal interesting insights into the human experience. At present, datasets used to train ML algorithms are typically benchmarked against human judgment.

- **Objective Metrics:** Quantitatively evaluating video quality according to technological parameters, objective metrics cover a wide variety of mathematical equations. Some metrics compare the unaltered and distorted versions of the video to determine an overall score, such as the peak signal-to-noise ratio (PSNR) and the Structural Similarity Index (SSIM). These measurements allow for fast, automated evaluations, but they are typically too simplistic to account for subtleties in human perception. For instance, PSNR is sensitive to inaccuracies but has a poor relationship with human perception.

- **Hybrid Approaches**: These methods integrate aspects of both subjective evaluation and more quantitative measures. The goal is to bring together scientific data and individual experience. To efficiently forecast quality, these approaches are trained on subjective input with the goal of factoring in human perception while preserving the efficiency gains from automation provided by objective measures.

- **Model-Based Techniques**: Computer simulations are used to mimic features of the human visual system to conduct quality assessments of software. Organizations such as the Video Quality Experts Group (VQEG) have created computational models like the Video Quality Model (VQM) and the Video Quality Expert (VQE) [57]. Contrast sensitivity and spatial masking are taken into account while ensuring improved consistency between technical measures and human perception.

- **Bitstream-Based Assessment**: In contrast to comprehensive decoding, these approaches use components such as encoding settings, frame rate, and compression ratios in the video bitstream. Although they are computationally efficient, these methods may miss certain subtleties in how content is perceived.

- **No-Reference (NR) Metrics**: Traditional VQA methods generally resorted to full-reference (FR) based techniques, which assess a distorted video against a reference, i.e., the original video, to gauge distortion. However, NR methods evaluate video quality without the need for a reference stream. This is also known as a reference-free metric. Video Quality Testing with Deep Learning (VQTDL) is an example of such an algorithm developed by TestDevLab that tries to evaluate quality without any external references, making it useful in situations where such resources are unavailable [58].

Since VQA is an application-centric process, the exact approach chosen needs to be highly adapted for the purpose. The techniques used around recent trends, as well as the need for VQA approaches to be adapted for emerging topics, are summarized in Figure 1-5 [59].

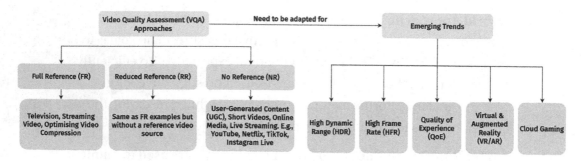

Figure 1-5. *Approaches and emerging trends for VQA*

Despite their historical significance, classic VQA techniques have major shortcomings when it comes to the abundance of today's multimedia content creation and deployment. Today's VQA methods are thus built around ML models that dynamically adapt to new video formats, varying content genres, and subtle perceptual differences. An overview of today's most adopted methodologies is given in Figure 1-6 [60].

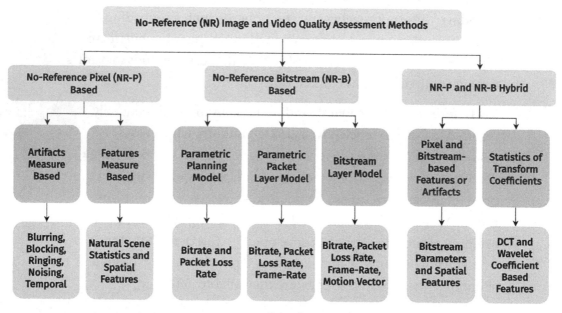

Figure 1-6. *No-reference image and video quality assessment methods*

1.2.4 Real-World Applications of VQA

The field of VQA is not limited to only scientific studies and technological audits. The ability to quantitatively assess and enhance video quality has several practical applications across a wide range of sectors. In this section, some of the practical settings in which VQA has proven to be particularly useful are investigated. They include the following:

- **Entertainment and Streaming Services**: Streaming platforms like Netflix, Amazon Prime Video, and Disney+ rely heavily on visual and audio quality for evaluating their QoE. Buffering and compression artifacts such as noisiness, blockiness, blurriness, and jerkiness are examples of video quality degradation that can be measured through constant monitoring. VQA is also used in content recommendation frameworks to suggest to users similar shows or movies that suit their likings based on past viewing history [61].

- **Telecommunications and Broadcasting**: In the case of telecom providers who often rely on IPTV and OTT services as their main source of revenue, service monitoring dashboards are crucial to have a bird's-eye view of all channels that they are broadcasting. This is particularly helpful in terms of proactively debugging streaming issues and reducing the number of client complaint calls that they register. Likewise, live events such as sports and news require a transmission certainty, which often decides if the end user will subscribe to the particular provider or not.

- **Health Care and Medical Imaging**: The reliability of medical diagnosis relies on high-quality visual data from a variety of imaging modalities, including X-rays, MRIs, and ultrasounds. To prevent incorrect diagnoses, VQA algorithms scour medical pictures for any signs of distortion or abnormalities, which alert medical experts [62]. The importance of maintaining high picture quality cannot be overstated in this field.

- **Education and E-Learning**: As a result of worldwide issues like the COVID-19 epidemic, there has been a dramatic increase in the use of e-learning tools in the education field, which to this day is still on the rise. High-quality educational video content is ensured through VQA algorithms that are present in video lectures, tutorials, interactive simulations, and video conferences encompassing large numbers of students to monitor and guarantee uninterrupted presentation.

- **Advertising and Marketing**: With increasing awareness about the relationship between video quality and target audience, marketers and advertisers use VQA to evaluate video quality before releasing video commercials, ensuring professionalism and dedication to the excellence of the brand in question, resulting in an increased likeliness to attract and retain viewer attention.

- **Security and Surveillance**: Live video feeds from surveillance cameras can be subject to strong VQA algorithms to ensure that recorded footage is not faulty due to compression or transmission flaws, allowing for accurate features such as car plate detection with AI.

- **Gaming and Virtual Reality (VR)**: Game development studios rely heavily on VQA to create truly immersive experiences for their players [63]. Gamers expect stunning visuals and flawless performance with the rise of high-end game development environments. VQA thus assists this proof with visual optimization and quality consistency.

These practical implementations illustrate the far-reaching effects of VQA on industry and daily life. VQA helps with user satisfaction, productivity, and even certain important decisions by guaranteeing the high-quality delivery of video material, which continues to rise to the top of the media food chain.

1.2.5 Challenges in VQA

Despite its significant strides, VQA encounters several persistent challenges, ranging from the subjective nature of human perception to the complexity of new video formats, thus requiring constant research and development. The very nature of the human mind makes perception very subjective and unique. What constitutes high-quality media to one viewer may not be the same to another. In the pursuit of objective metrics for quality evaluation, VQA faces a fundamental hurdle in the form of subjectivity. Constant effort is required to close the "subjectivity gap" by developing models that reflect the wide variety of human perspectives [64]. Likewise, videos are ever-evolving works that include shifting scene compositions, kinetic elements, and auditory cues. Temporal characteristics, such as motion smoothness, synchronization, and consistency, have a major influence on viewer pleasure but are difficult to analyze for VQA. Moreover, the need for VQA to be both scalable and ubiquitous is increasingly becoming an urgent issue with the imminent explosion of user-generated material on applications like YouTube, TikTok, Reddit, and Instagram. Assessing video quality across massive datasets in multiple environments, from all kinds of sources, in a timely and reliable manner is an ongoing challenge. This also resonates with media being uploaded from different devices of varying capabilities, thus with different video formats and codecs, ranging from 4K video to HDR, to 360-degree VR material. VQA has to evolve continuously to assess quality in these newer forms [65].

One of the main impediments, however, remains the requirement that VQA algorithms be carried out in real time for intensive applications such as video conferencing and live streaming, which encompass massive numbers of clients ranging from a few hundred on Instagram Lives to several thousand on gaming streams on Twitch. This can prove to be computationally demanding, and may not work as planned in low-resource environments such as PCs with low specifications. On the same wavelength, there is a lack of ground-truth data, where datasets need to be built independently around each platform, making subjective data collection a time-consuming and costly factor. Network circumstances are also taken into account in some algorithms when measuring video quality. This is present in adaptive streaming protocols such as Dynamic Adaptive Streaming over HTTP (DASH). Since content quality might change mid-stream, VQA must be flexible so as to accurately evaluate video streams of an evolving nature in adaptive streaming settings. These issues can be tackled with advancements in computer vision, ML, psychology, and data science and require collaboration between researchers, content creators, and industry stakeholders to enhance VQA approaches and guarantee its functioning in this rapidly expanding digital realm.

1.2.6 Emerging Trends in VQA

This section caters to the recent trends and developments in the VQA sphere, with the ever-changing digital content era and the rising significance of quality monitoring [66, 67, 68].

- **Deep Learning Advancements**: Developments in DL have allowed VQA to reach previously unattained levels with algorithms such as convolutional neural networks (CNNs) and recurrent neural networks (RNNs). Sophisticated subjective features of video quality are increasingly being modeled using state-of-the-art DL architectures. These models are trained on large datasets of human judgments of quality and can extract complex visual elements from video to make reliable judgments.

- **No-Reference (NR) Metrics**: Metrics that don't rely on objective and standardized examples of video quality have recently come into vogue, and for good reason. Metrics like these, which are often based on ML methods, may be especially helpful in cases where reference movies are either not accessible or are difficult to get.

- **Multi-Modal Assessment**: Integrating sound, texts, and metadata into a single assessment offers a new direction. Assessment models can gain a deeper knowledge of video quality since they take into account different modalities. For instance, a comprehensive evaluation of a multimedia experience may be attained by considering both the visual and audio components.

- **Quality-of-Experience (QoE) Models**: These frameworks take into account user-centric characteristics beyond those measured by standard quality measurements. User participation, focus, and reaction are all considered, making these models crucial for enhancing video in ways that appeal to the widest possible audience.

- **Adaptive Streaming Assessment**: The prevalence of adaptive streaming technologies has resulted in newer algorithms to reduce noticeable breaks in video quality as they adapt to changing network environments, with the incorporation of NTMA.

- **Explainable AI (XAI)**: Explainable AI is more pronounced in the health-care and security realms, where the model provides the reasoning behind a particular model's quality assessment, further bolstering evaluation accuracy.

- **Edge Computing for Real-Time VQA**: As one of the methods to provide real-time analysis, edge computing is being used to analyze data closer to the source, such as the client device. A typical example includes a live-streaming or video conferencing setting where edge-based VQA systems are built around the user equipment to achieve low latency and fast quality evaluation.

- **Cross-Domain Applications**: Algorithms developed in one field are currently being transposed and adjusted for others. For example, techniques developed for streaming media are being increasingly used in the medicine, education, and security sectors for high-quality images, VR classrooms, and camera feeds, respectively.

These developments are proof that VQA is becoming a fundamental part of the ecosystem that supports online media. To meet the growing need for elevated, interactive experiences in a variety of disciplines, VQA will play a prominent role as internet video continues to dominate communications, entertainment, and information distribution.

1.3 Machine Learning in JavaScript

Machine learning has transformed numerous industries by allowing systems to learn from observations and enhance performance over time. With JavaScript's ascent to prominence as a flexible and widely used programming language, it is now possible to incorporate machine learning algorithms directly in web applications. This section delves into how JavaScript is used for machine learning tasks, providing powerful tools and frameworks that add advanced analytics and intelligent capabilities to web-based environments.

1.3.1 Introduction to Machine Learning

The revolutionary unit of artificial intelligence (AI) known as machine learning (ML) allows machines to learn from data and make predictions or choices without being explicitly programmed to do so. The advent of ML support in JavaScript over the last few years has unleashed incredible potential for those working in web development and engineering. This section introduces the basics of ML and explains why it matters in the JavaScript context.

Machine learning is fundamentally concerned with generalization and pattern identification. It entails teaching computers to analyze large amounts of data and draw meaningful conclusions or make accurate forecasts. Machine learning algorithms may analyze past information to determine patterns and generate predictions about unanticipated data. ML can be broken down into the following primary categories [69]:

1. **Supervised Learning**: Supervised learning involves training models using labeled data that contains both the input and the expected output. The objective is to figure out how to map inputs to outputs. Detecting spam emails, classifying images, and regression problems are some examples.

2. **Unsupervised Learning**: Unsupervised learning works with unlabeled data, and the algorithm attempts to discover concealed structures or patterns in the data. This includes clustering and dimensionality reduction, which may be used for market segmentation or identifying anomalies.

3. **Reinforcement Learning**: In this type of learning, agents are equipped with the ability to make decisions in an environment to maximize a reward signal. Some use cases are robotics, gaming, and autonomous systems.

Given its prevalence in web development, JavaScript's position in ML has grown in significance, enabling developers to incorporate lightweight and sophisticated algorithms directly into web apps. This results in engaging and adaptive user experiences. JavaScript tools and frameworks have evolved to make ML readily available to a larger audience. This also opens the door to real-time possibilities that do not require server-side computation. What may seem like complex tasks, such as object detection in photos, natural language processing, sentiment analysis, and forecasting analytics, are natively packaged and built as ML models at the client side, right in the browser.

Moreover, with a constantly evolving set of tools that also have open-source involvement from worldwide contributors, JavaScript is heavily infused with libraries such as TensorFlow.js, Brain.js, and ml5.js, which provide data preparation methods as well as abstractions and functions for pre-trained models [70]. These frameworks provide the average developer with an edge for delving into advanced ML techniques mapped onto simple methods that only require data in the correct format. Additionally, neural networks can also be built and tailored to match the requirements of datasets. As a result, image recognition, chatbots, and predictive analytics are readily available to the developer. On the same wavelength, JavaScript allows code writers to use ML in areas such as NTMA and VQA with off-the-shelf classes and methods that come with the available libraries. The coupling of JavaScript with ML heralds a new age of web-based applications that are not solely adaptive but also smart and geared toward making data-driven decisions instantaneously.

1.3.2 Coupling JavaScript with Machine Learning

The amalgamation of JavaScript and ML renders application development much easier, with built-in methods for data preparation and analytics. This section focuses on the gap bridged between web development and data-driven intelligence [71], as follows:

- **Web-Driven Integration**: For several decades, JavaScript has been the dominant language for frontend web design. Its combination with ML adds intelligence to web apps directly, allowing them to better interpret, evaluate, and react to user engagements. This connectivity is extremely valuable for developing dynamic and customized user experiences.

- **Client-Side Machine Learning**: JavaScript gives programmers the ability to house and apply ML models natively at the client side; i.e., inside web browsers. This is a game changer, since it enables real-time use of the device's processing capabilities, without needing constant server interactions. This means that ML-powered applications can also be used offline or in cases of poor network connectivity.

- **Interactive Data Visualization**: In the ML realm, data visualization is usually regarded as one of the basic requirements of the application being built. With most of the use cases in this space being real time, JavaScript offers visualization libraries such as D3.JS to complement ML workflows. This means that by adding to the forecast produced by the algorithm, the data streams can be conveyed to the end user in terms of interactive graphs and charts, together with the result of the framework, further adding a layer of comfort for the developer when building ML-driven applications.

- **Cross-Platform Compatibility**: Since JavaScript is inherently cross-platform compatible, ML-powered web apps may operate unaltered across a range of hardware and operating systems. By allowing users to access ML apps from smartphones, tablets, desktops, and other devices, the popularity of ML applications is enhanced.

- **Real-Time Interactivity**: The real-time interactivity requirement of today's applications goes hand in hand with JavaScript's event-driven architecture, which is ideal for query–response interactions in ML applications such as chatbots, sentiment analysis, and live data visualization.

- **Enhancing User Experiences**: Adaptive user interfaces that change based on user actions and preferences are feasible with the help of ML supported by JavaScript. This new level of sophistication can suggest tailored content and offer context-aware recommendations.

The practical implementations of ML in JavaScript result in a powerful mix that can be used to create applications that are not only visually appealing, but also intelligent, event driven, and revolutionary in the realm of web development.

1.3.3 Data Preparation and Preprocessing in JavaScript

Accurately preparing and cleaning data is a major element of the machine learning workflow. This section delves into how JavaScript assists with these critical stages, getting data prepared for ML model training and inference. Table 1-2 summarizes the available features [72, 73, 74].

Table 1-2. *Data Preparation and Preprocessing Features in JavaScript*

Feature	Description
Data Handling Libraries	JavaScript includes packages that make data operations and preparation easier. Pandas.js and Dataframe-js are examples of libraries that offer data structures and methods comparable to Python's Pandas, enabling developers to effectively import, filter, and manipulate data.
Data Acquisition	It is typical for web applications to obtain content from APIs, databases, and user contributions. Some libraries such as Axios ease the process of HTTP queries for data retrieval, while IndexedDB offers a client-side database for storage and requests for web applications.
Data Cleaning	Real-world data often contains discrepancies, missing numbers, and outliers, requiring a cleaning step before feeding the training algorithm. Frameworks such as Lodash, D3.js, and Dataframe-js facilitate this process by eliminating duplicates, filling in missing values, and identifying anomalies.
Feature Engineering	Custom functions in JavaScript with regard to ML also allow the process of developing new features from existing data to enhance model performance. This can be based on timestamps, text data, or geographical data. For time-related manipulations, Moment.js can be used, while "Natural" and "Compromise" are examples of natural processing language (NLP) libraries. Some geospatial frameworks include turf.js and leaflet.js.
Scaling and Normalization	Proper scaling and normalization methods can be created through custom methods using the Math.js library, allowing ML algorithms to converge better.
Data Splitting	When it comes to evaluating model performance, data is often partitioned into training and testing datasets. Certain libraries have methods for data manipulation that enable random or stratified data splitting, thus eradicating bias. For example, the scikit.js has an in-built function that enables splitting a dataset into training and testing ratios either randomly or with specified proportions.
Data Visualization	As one of the prerequisites of ML coupled with dashboards, visualization tools are fundamental for user understanding. Libraries like D3.js and Chart.js bring the user an edge closer to interactivity and the ability to explore predicted or classified data instantaneously.

(continued)

Table 1-2. (*continued*)

Feature	Description
Handling Time Series Data	Certain applications built around stock prices and IoT sensors can largely benefit from functions such as resampling, aggregating, and interpolating time-series data points.

Often underestimated, data preparation tools pave the way for the creation of accurate ML models through effective management, cleaning, and data transformation.

1.3.4 Supervised Learning with JavaScript

One of the pillars of ML is the supervised learning approach, which allows frameworks to be trained for forecasting and classification using labeled data. This method gives developers a robust set of resources for making smart websites. The following is a list of refined points of how the combination of ML with JavaScript enables programmers to efficiently create supervised learning models [75]:

- **Leveraging Labeled Data**: JavaScript is versatile for applications like sentiment analysis, content categorization, and recommendation engines since it allows for the gathering of labeled information directly through web interfaces. This means that inputs are linked to desired outcomes.

- **Streamlined Model Training**: Coders may use straightforward syntax to build model architectures, set hyperparameters, and kick off training. The real-time nature of JavaScript allows for the visualization of training progress, increasing the openness of the model and allowing for the possibility of dynamic modifications.

- **Cross-Platform Compatibility**: Such models built with JavaScript work reliably on a wide range of devices and operating systems, indicating that a wide audience may use such online applications that have ML capabilities, regardless of the device type.

- **Applications Across Domains**: Natural language processing (NLP), image recognition, fraud detection, AI chat bots, and other applications can be developed, which opens the door to cutting-edge, highly adaptable online apps that can meet the demands of virtually everyone.

Some use cases of supervised learning are as follows:

- **Classification**: The categorization of texts according to their subjects or emotions is a typical application of supervised learning. JavaScript may be used to create an email filter that determines if a message is spam or not based on training data.

- **Image Recognition**: Models for classifying photos, such as determining whether or not a given picture depicts a cat or a dog, can be implemented in JavaScript.

- **Predictive Analytics**: The prediction of stock prices, customer turnover, and client demand can be performed dynamically through websites crafted with JavaScript, using tools like scikit-learn.js for regression and time-series analysis.

1.3.5 Unsupervised Learning with JavaScript

Unsupervised learning thrives on detecting hidden structures, patterns, and correlations in unlabeled datasets as compared with labeled ones in supervised learning. This allows developers to acquire insights and extract meaningful information from unannotated data. The following are notable features [76]:

- **Clustering and Data Grouping**: Data points can be grouped based on underlying similarities through what is known as data clustering. This proves to be highly useful in cases such as client segmentation for customized online experiences.

- **Dimensionality Reduction**: To reduce the computing complexity in web applications, techniques such as principal component analysis (PCA) and t-distributed stochastic neighbor embedding (t-SNE), which specialize in reducing the dimensionality of datasets, are used.

- **Anomaly Detection**: Unsupervised learning excels in spotting abnormalities or outliers in data trends, which is crucial in the fraud detection and network security sphere. Deviations from the norm are quickly identified and resolved accordingly.

- **Topic Modeling in Natural Language Processing (NLP)**: Having gained massive popularity in recent years, JavaScript libraries like Natural.js empower programmers to infer meaning and group similar words from large unstructured text data, in turn facilitating content categorization and recommendation. This is an increasingly important feature that is used by online shops to gain insights about reviews on particular products.

Some use cases of unsupervised learning are as follows:

- **Clustering**: Algorithms such as K-means clustering or hierarchical clustering can be used to break down customers into different groups according to their online behaviors or product preferences.

- **Anomaly Detection**: Engines revolving around finding odd patterns in financial data to identify fraudulent transactions can be built.

- **Dimensionality Reduction**: As mentioned previously, approaches like PCA can be used to reduce data dimensionality while keeping the crucial fields.

Coupled with web development in JavaScript, unsupervised learning is a powerful ally that reveals hidden patterns in data, deconstructs complicated datasets, and extracts insightful knowledge. Regardless of the availability of labeled data, developers can create smart and data-informed online experiences through correlations and inference.

1.3.6 Deep Learning in JavaScript

As a major subset of ML, deep learning (DL) has skyrocketed in recent years due to its capacity to solve intricate problems using neural networks with numerous layers. DL is mostly linked with languages like Python through frameworks such as TensorFlow and PyTorch, but it has also seeped through the JavaScript realm, with a solid foundation provided to developers through the use of Node.js libraries. DL leverages the potential of artificial neural networks (ANNs), which are meant to imitate the structure and operation of the human brain, in turn adopting the same thinking strategy. Just like the brain is made up of neurons that trigger responses, these networks are driven by hyperparameters such as weights and activation functions that help them learn hierarchical characteristics and representations from inputs. After being trained repeatedly over a series of different datasets, the algorithm can process and yield outputs for intricate tasks. The rise of DL in the web development sphere is due to the following factors [77]:

- **Versatile Neural Network Architectures**: JavaScript, together with packages such as TensorFlow.js, provides extensive support for a diverse range of neural network topologies. Convolutional neural networks (CNNs) have exceptional performance in the domain of image analysis, whereas recurrent neural networks (RNNs) are specifically tailored to handle sequential data, such as natural language processing (NLP) [78]. This field is being significantly impacted by the emergence of advanced transformer models such as BERT and GPT-3, which are bringing about a paradigm shift in the realms of language comprehension and generation [79]. The flexibility of web applications enables them to effectively use DL techniques for various purposes, such as image recognition, sentiment analysis, and language translation.

- **Transfer Learning with Pre-Trained Models**: The deep learning environment in JavaScript enables the practice of transfer learning, wherein developers can refine pre-trained models to suit specific applications. For example, employing a pre-trained CNN such as VGG16 as a feature extractor in the context of image recognition can significantly accelerate the process of software development [80]. In a similar vein, the use of BERT's situational language comprehension capabilities serves to augment the precision of chatbots and the summary of material in activities related to NLP.

- **Real-Time Inference with JavaScript**: One significant benefit of implementing DL with JavaScript is its capacity to conduct instantaneous inference directly inside web browsers. The use of client-side processing serves to mitigate latency and diminish the necessity for server exchanges. Algorithms like EfficientNet, designed for image sorting, and BERT, developed for textual comprehension, can deliver prompt responses to user inputs [81]. This is renowned in applications such as chatbots, recommendation systems, and content personalization.

- **Interpretability and Model Explainability**: Familiarity with the decision-making processes of DL models becomes increasingly important as they grow in complexity. JavaScript is equipped with tools to interpret models, enabling them to visually represent the significance of features and decision boundaries. Methods such as integrated gradients and SHapley Additive exPlanations (SHAP) values provide insight into the rationale behind a mode's prediction, hence enhancing transparency and reliability in fields such as health-care diagnostics and credit risk assessment [82].

- **Cross-Domain Impact**: The application of deep learning in JavaScript has permeated various sectors, exerting a wide-ranging impact. CNNs are employed in the health-care industry to examine medical images for diagnosis. Conversely, in the realm of self-driving cars, algorithms like YOLO (You Only Look Once) are used to identify obstacles in real time [83]. Likewise, when it comes to NLP, transformer algorithms such as BERT and GPT-3 have brought about a significant transformation in the capabilities of chatbots and language understanding.

The integration of DL into JavaScript offers web developers a diverse range of technological functionalities, facilitating the creation of intelligent, dynamic, and swift applications that demonstrate exceptional performance. This meets the increasing demands of internet users for superior performance to go along with the rising use of web applications.

1.3.7 Deploying Machine Learning Models in Web Applications

The incorporation of machine learning models into web-based applications signifies the completion of development powered by data, enabling consumers to access intelligence and automation directly. This section delves into the fundamental components of implementing machine learning models in web applications, with a focus on practical challenges and opportunities [84], including the following:

- **Integration with Web Environments**: The ability of JavaScript to offer both server-side and client-side scripting allows a versatile platform with smooth integration.

- **RESTful APIs and Endpoints**: A frequently employed strategy for deploying models involves the development of RESTful APIs and corresponding endpoints. These application programming interfaces (APIs) provide a means for web applications to connect to and use ML models. This allows them to transmit data to make predictions and receive insights that are provided by the models. JavaScript, when used in conjunction with frameworks such as Express.js, facilitates the process of developing and managing APIs.

- **Client-Side Inference**: Frameworks such as TensorFlow.js facilitate the execution of client-side code. This approach decreases the amount of time it takes for data to be transmitted and reduces the frequency of interactions with the server, resulting in a user experience that is prompt and more efficient.

- **Scalability and Resource Management**: With frequent connectivity issues such as user load times and the number of simultaneous connections, scalability is crucial when deploying ML models in web browsers. Node.js offers scalability alternatives for effectively handling augmented traffic and ensuring quick execution of models.

- **Performance Optimization**: The optimization of model efficiency within web applications is of paramount significance. With asynchronous properties, JavaScript has the potential to optimize the framework execution, thus ensuring efficient generation of predictions, even in situations characterized by substantial user traffic.

1.4 Node.js and Networking

Node.js is a JavaScript runtime powered by Chrome's V8 JavaScript engine, which has garnered substantial recognition within the web development and networking realm. The fundamental basis for Node.js' suitability for networking is its use of an asynchronous, non-blocking input/output model and event-driven architectural design [85]. In contrast to conventional synchronous programming languages, Node.js can concurrently process many network requests without impeding the performance of other processes. This feature greatly enhances the efficiency of managing network connections, particularly in scenarios that prioritize responsiveness and real-time interactions [86]. In addition to allowing the development of complete networking applications at the server side, Node.js is cross-platform and runs on Windows, macOS, and several Unix-based operating systems. This is suited for installing these servers on an array of devices, ranging from web servers to IoT devices.

Node.js is known for its extensive collection of modules and libraries that can be accessed via its package management, "npm." A significant number of these modules are specifically designed for networking purposes, providing ready-made solutions for commonly encountered obstacles, such as the establishment of HTTP servers, implementation of WebSockets, and connecting with databases. The extensibility of the system facilitates the expeditious progress of development and streamlines the incorporation of third-party services into networking applications. The event-driven nature of Node.js offers significant benefits in situations involving real-time communication. WebSockets serves as a fundamental component of multiple applications, facilitating the establishment of bidirectional and low-latency communication channels between clients and servers [87]. The capacity to provide real-time updates is crucial in applications that necessitate immediate information dissemination, such as chat apps, online gaming, and collaboration tools.

The relationship between Node.js and networking is intertwined, as Node.js provides a robust framework for developing efficient and responsive networking applications. In this book, the possibilities of using Node.js to power up networking functionalities will be explored. These include the creation of servers and clients, ensuring the security of network connections, and facilitating the scalability of applications by catering to multiple simultaneous connections. Some use cases of Node.js in networking are listed in Table 1-3 [88, 89].

Table 1-3. *Use Cases of Node.js in Networking*

Use Case	Description
E-commerce and Real-Time Inventory Management	Node.js empowers e-commerce platforms to deliver real-time stock updates, in turn guaranteeing a smooth and uninterrupted shopping experience for their clients.
Collaborative Tools and Chat Applications	As one of the prerequisites of chat applications, instantaneous two-way communication among users is provided by Node.js libraries such as Socket.io, which are highly functional and interactive.
Streaming and Media Services	Video playback and content distribution are guaranteed by servers coded with low-latency and high-throughput capabilities.
Social Media and Real-Time Updates	To maintain user engagement, Node.js can facilitate the provision of push notifications, real-time updates, and instant chat capabilities, thus promoting user involvement and attention retention.
Online Gaming and Multiplayer Platforms	Online gaming systems require efficient and immediate communication capabilities, such that real-time interactions among several players across the world are supported in a matter of milliseconds. Server-side development of frameworks facilitates real-time gameplay and synchronization.
Health Care and Telemedicine	Secure and real-time communication is a fundamental requirement for the effective functioning of telemedicine applications, bridging the gap between health-care providers and patients. These features are ensured by Node.js, which can be employed in the development of telehealth systems, including video consultations, remote monitoring, and the management of patient data.
Real-Time Analytics and Dashboards	Business decisions revolve around website traffic, user behavior, and application performance, all possible with the conjunction of Node.js.

1.5 Book Overview

This book is organized into seven chapters. In the first, i.e., the current chapter, the groundwork has been laid for exploration. An overview of network traffic monitoring and analysis (NTMA) and video quality assessment (VQA) has been provided.

The core concepts, methodologies, and existing applications of machine learning (ML) in JavaScript are also elaborated, together with the development of networking applications using Node.js. These topics introduce the reader to the essential components of these paradigms.

Chapter 2 focuses on the machine learning techniques in the sphere of NTMA. It introduces the fundamentals of network traffic analysis in addition to current applications being used by industry professionals. A detailed state-of-the-art review of the use of NTMA with regard to research and applications in the real world is also given.

Chapter 3 mirrors this structure for VQA by setting the focus on the fundamental concepts involved in producing a quality assessment metric through both subjective and objective assessment.

Chapter 4 shifts the spotlight to a breakdown of classification and prediction algorithms revolving around NTMA. This involves a detailed examination of ML models and frameworks, together with their hyperparameters, for optimal accuracy. Additionally, the application of support vector machine (SVM) and deep learning (DL) methodologies in the context of VQA are also unraveled.

Chapter 5 delves into the practical aspects of implementing NTMA with JavaScript, starting with how network traffic is captured, before being subject to a real-time network analysis through Node.js's backend. The application was packaged as a lightweight browser extension that could be opened on any page for immediate real-time monitoring.

Chapter 6 describes the development of the VQA application, whose backend is built using Java and hosted on a local servlet through Apache Tomcat. With the frontend comprising the browser extension opened on a browser tab playing a video, the feed is sent to the servlet, in turn triggering the quality assessment block. This yields a video-quality score.

Chapter 7 combines the features of NTMA and VQA into a single application where both capabilities are available under a single browser extension and at the click of a button. The system model, program structures, and performance analysis aspects are all covered.

1.6 References – Chapter 1

1.   T. P. Fowdur, B. N. Baulum, and Y. Beeharry, "Performance Analysis of Network Traffic Capture Tools and Machine Learning Algorithms for the Classification of Applications, States and Anomalies," *International Journal of Information Technology*, vol. 12, no. 3, pp. 805–24, Apr. 2020, doi: 10.1007/s41870-020-00458-0.

2.   C. S. Inc, "Cisco Predicts More IP Traffic in the Next Five Years Than in the History of the Internet," GlobeNewswire News Room, Nov. 27, 2018, https://www.globenewswire.com/en/news-release/2018/11/27/1657381/0/en/Cisco-Predicts-More-IP-Traffic-in-the-Next-Five-Years-Than-in-the-History-of-the-Internet.html (accessed Aug. 28, 2023).

3.   MantisNet, "Network Traffic Analysis: Real-time Identification, Detection and Response to Threats," https://www.mantisnet.com/blog/network-traffic-analysis (accessed Aug. 28, 2023).

4.   K. Kashyap, "Strategy and OKRs Are Key to Successful AI Adoption," Spiceworks, May 31, 2023. https://www.spiceworks.com/hr/hr-strategy/articles/strategy-and-okrs-key-to-successful-ai-adoption/ (accessed Aug. 28, 2023).

5.   A. Patrizio, "IDC: Expect 175 Zettabytes of Data Worldwide by 2025," Network World, Dec. 03, 2018, https://www.networkworld.com/article/3325397/idc-expect-175-zettabytes-of-data-worldwide-by-2025.html (accessed Aug. 28, 2023).

6.   P. Joshi, A. Bhandari, K. Jamunkar, K. Warghade, and P. Lokhande, "IJARCCE Network Traffic Analysis Measurement and Classification Using Hadoop," *International Journal of Advanced Research in Computer and Communication Engineering*, vol. 5, no. 3, 2016, doi: https://doi.org/10.17148/IJARCCE.2016.5360.

7.   N. James, "45 Global DDOS Attack Statistics 2023 - Astra Security Blog," Getastra, Dec. 27, 2022, https://www.getastra.com/blog/security-audit/ddos-attack-statistics/ (accessed Sep. 17, 2023).

8.   Statista, "Global Digital Population 2022," Statista, Apr. 03, 2023, https://www.statista.com/statistics/617136/digital-population-worldwide/ (accessed Sep. 17, 2023).

9.   A. D'Alconzo, I. Drago, A. Morichetta, M. Mellia, and P. Casas, "A Survey on Big Data for Network Traffic Monitoring and Analysis," *IEEE Transactions on Network and Service Management*, vol. 16, no. 3, pp. 800–13, Sep. 2019, doi: https://doi.org/10.1109/tnsm.2019.2933358.

10. T. Kellermann and R. McElroy, "Modern Bank Heists 4.0," `https://www.vmware.com/content/dam/digitalmarketing/vmware/en/pdf/docs/vmwcb-report-modern-bank-heists-2021.pdf` (accessed Sep. 17, 2023).

11. N. Tiushka, "60+ Internet Speed Statistics: Emerging Trends, The Impact Of 5G/6G And Internet Availability & Use," *MarketSplash*, Aug. 04, 2023.

12. F. Pacheco, E. Exposito, M. Gineste, C. Baudoin, and J. Aguilar, "Towards the Deployment of Machine Learning Solutions in Network Traffic Classification: A Systematic Survey," *IEEE Communications Surveys & Tutorials*, vol. 21, no. 2, pp. 1988–2014, 2019, doi: `https://doi.org/10.1109/comst.2018.2883147`.

13. D. Reinsel, J. Gantz, and J. Rydning, "The Digitization of the World from Edge to Core," Nov. 2018, `https://www.seagate.com/files/www-content/our-story/trends/files/idc-seagate-dataage-whitepaper.pdf` (accessed Sep. 17, 2023).

14. L. Vailshery, "IoT Connected Devices Worldwide 2019–2030," Statista, Jun. 2023, `https://www.statista.com/statistics/1183457/iot-connected-devices-worldwide/` (accessed Sep. 17, 2023).

15. Z. Trabelsi, H. E. Sayed, L. Frikha, and T. Rabie, "A Novel Covert Channel Based on the IP Header Record Route Option," *International Journal of Advanced Media and Communication*, vol. 1, no. 4, p. 328, 2007, doi: `https://doi.org/10.1504/ijamc.2007.014811`.

16. V. Jain, *Wireshark Fundamentals: A Network Engineer's Handbook to Analyzing Network Traffic*, 2022, doi: `https://doi.org/10.1007/978-1-4842-8002-7`.

17. "VNI Complete Forecast Highlights Global Internet Users: % of Population Devices and Connections per Capita Average Speeds Average Traffic per Capita per Month Global: 2021 Forecast Highlights IP Traffic," 2016, `https://www.cisco.com/c/dam/m/en_us/solutions/service-provider/vni-forecast-highlights/pdf/Global_2021_Forecast_Highlights.pdf` (accessed Sep. 20, 2023).

18. Sandvine, "How 'App QoE' Can Increase Profitability While Improving Subscriber Satisfaction," 2023, `https://www.sandvine.com/hubfs/Sandvine_Redesign_2019/Downloads/2023/PDF/BRO-Sandvine-FEB2023.pdf` (accessed Nov. 05, 2023).

19. Sandvine, "Phenomena," 2023, `https://www.sandvine.com/hubfs/Sandvine_Redesign_2019/Downloads/2023/reports/Sandvine%20GIPR%202023.pdf` (accessed Nov. 05, 2023).

20. A. Siswanto, A. Syukur, E. A. Kadir, and Suratin, "Network Traffic Monitoring and Analysis Using Packet Sniffer," IEEE Xplore, Apr. 01, 2019, `https://ieeexplore.ieee.org/document/8742369` (accessed Sep. 20, 2023).

21. J. Koumar, K. Hynek, and Tomáš Čejka, "Network Traffic Classification Based on Single Flow Time Series Analysis," arXiv (Cornell University), Jul. 2023, doi: `https://doi.org/10.48550/arxiv.2307.13434`.

22. D. Mistry, P. Modi, K. Deokule, A. Patel, H. Patki, and O. Abuzaghleh, "Network Traffic Measurement and Analysis," IEEE Xplore, Apr. 01, 2016, `https://ieeexplore.ieee.org/abstract/document/7494141` (accessed May 26, 2021).

23. M. Uma and G. Padmavathi, "An Efficient Network Traffic Monitoring for Wireless Networks," *International Journal of Computer Applications*, vol. 53, no. 9, pp. 51–9, Sep. 2012, doi: `https://doi.org/10.5120/8453-2255`.

24. F. B. Insights, "Network Monitoring Market to Surpass USD 6.97 Billion by 2030, exhibiting a CAGR of 11.1%," GlobeNewswire News Room, Sep. 18, 2023, `https://www.globenewswire.com/en/news-release/2023/09/18/2744576/0/en/Network-Monitoring-Market-to-Surpass-USD-6-97-Billion-by-2030-exhibiting-a-CAGR-of-11-1.html` (accessed Sep. 20, 2023).

25. B. Kurt, E. Zeydan, U. Yabas, I. A. Karatepe, G. K. Kurt, and A. T. Cemgil, "A Network Monitoring System for High Speed Network Traffic," 2016 13th Annual IEEE International Conference on Sensing, Communication, and Networking (SECON), Jun. 2016, doi: `https://doi.org/10.1109/sahcn.2016.7732965`.

26. Y. A. Al-Sbou, "Wireless Networks Performance Monitoring Based on Passive-Active Quality of Service Measurements," *International Journal of Computer Networks & Communications*, vol. 12, no. 6, pp. 15–32, Nov. 2020, doi: `https://doi.org/10.5121/ijcnc.2020.12602`.

27. B. Yang and D. Liu, "Research on Network Traffic Identification Based on Machine Learning and Deep Packet Inspection," IEEE Xplore, Mar. 01, 2019, `https://ieeexplore.ieee.org/document/8729153` (accessed Sep. 20, 2023).

28. J. David and C. Thomas, "Intrusion Detection Using Flow-Based Analysis of Network Traffic," *Communications in Computer and Information Science*, Jan. 2011, doi: `https://doi.org/10.1007/978-3-642-17878-8_40`.

29. V. Stavrinides, "Integrating Cyber Security into the ESG Narrative," TORI, `https://www.toriglobal.com/blog/integrating-cyber-security-into-the-esg-narrative/` (accessed Sep. 20, 2023).

30. J. Cao, D. Wang, Z. Qu, H. Sun, B. Li, and C.-L. Chen, "An Improved Network Traffic Classification Model Based on a Support Vector Machine," *Symmetry*, vol. 12, no. 2, p. 301, Feb. 2020, doi: `https://doi.org/10.3390/sym12020301`.

31. A. Shahraki and Ø. Haugen, "An Outlier Detection Method to Improve Gathered Datasets for Network Behavior Analysis in IoT," *Journal of Communications*, vol. 14, no. 6, pp. 455–62, 2019, doi: https://doi.org/10.12720/jcm.14.6.455-462.

32. A. Shahraki, M. Geitle, and Ø. Haugen, "A Comparative Node Evaluation Model for Highly Heterogeneous Massive-Scale Internet of Things-Mist Networks," *Transactions on Emerging Telecommunications Technologies*, Mar. 2020, doi: https://doi.org/10.1002/ett.3924.

33. M. Abbasi, A. Shahraki, and A. Taherkordi, "Deep Learning for Network Traffic Monitoring and Analysis (NTMA): A Survey," *Computer Communications*, vol. 170, pp. 19–41, 2021, doi: https://doi.org/10.1016/j.comcom.2021.01.021.

34. Z. Wang, K. W. Fok, and V. L. L. Thing, "Machine Learning for Encrypted Malicious Traffic Detection: Approaches, Datasets and Comparative Study," *Computers & Security*, vol. 113, p. 102542, Feb. 2022, doi: https://doi.org/10.1016/j.cose.2021.102542.

35. Z. Cheng, M. Beshley, H. Beshley, O. Kochan, and O. Urikova, "Development of Deep Packet Inspection System for Network Traffic Analysis and Intrusion Detection," IEEE Xplore, Feb. 01, 2020, https://ieeexplore.ieee.org/document/9088535 (accessed Aug. 09, 2021).

36. "The Growth in Connected IoT Devices Is Expected to Generate 79.4ZB of Data in 2025, According to a New IDC Forecast," www.businesswire.com, Jun. 18, 2019, https://www.businesswire.com/news/home/20190618005012/en/The-Growth-in-Connected-IoT-Devices-is-Expected-to-Generate-79.4ZB-of-Data-in-2025-According-to-a-New-IDC-Forecast (accessed Sep. 23, 2023).

37. C. Ene, "Council Post: 10.5 Trillion Reasons Why We Need a United Response to Cyber Risk," *Forbes*, Feb. 22, 2023. https://www.forbes.com/sites/forbestechcouncil/2023/02/22/105-trillion-reasons-why-we-need-a-united-response-to-cyber-risk/?sh=c2991fd3b0c4 (accessed Sep. 23, 2023).

38. K. Alexis Fidele, Suryono, and W. Amien Syafei, "Denial of Service (DoS) Attack Identification and Analyse Using Sniffing Technique in the Network Environment," *E3S Web of Conferences*, vol. 202, no. 15003, 2020, doi: https://doi.org/10.1051/e3sconf/202020215003.

39. T. Randhawa and A. Haque, "NIMA (Network Impact Modeling and Analysis): A QoS Perspective," GLOBECOM 2020 - 2020 IEEE Global Communications Conference, Dec. 2020, doi: https://doi.org/10.1109/globecom42002.2020.9348277.

40. M. Patzold, "5G Is Going Live in Country After Country [Mobile Radio]," *IEEE Vehicular Technology Magazine*, vol. 14, no. 4, pp. 4–10, Dec. 2019, doi: `https://doi.org/10.1109/mvt.2019.2939756`.

41. D. A. Keim, F. Mansmann, J. Schneidewind, and T. Schreck, "Monitoring Network Traffic with Radial Traffic Analyzer," 2006 IEEE Symposium on Visual Analytics Science And Technology, Oct. 2006, doi: `https://doi.org/10.1109/vast.2006.261438`.

42. "Network Operation Analytics," Siemens, `https://www.siemens.com/global/en/products/services/digital-enterprise-services/analytics-artificial-intelligence-services/network-operation-analytics.html` (accessed Sep. 23, 2023).

43. Tulsi Pawan Fowdur and L. Babooram, *Real-Time Cloud Computing and Machine Learning Applications*, New York: Nova Science, 2021.

44. "Cisco ThousandEyes End User Monitoring," Cisco, `https://www.cisco.com/c/en/us/products/cloud-systems-management/thousandeyes-end-user-monitoring/index.html` (accessed Sep. 23, 2023).

45. T. P. Fowdur and L. Babooram, "Performance Analysis of a Cloud-Based Network Analytics System with Multiple-Source Data Aggregation," *International Journal of Pervasive Computing and Communications*, Sep. 2022, doi: `https://doi.org/10.1108/ijpcc-06-2022-0244`.

46. E. W. Biersack, C. Callegari, and Maja Matijasevic, *Data Traffic Monitoring and Analysis*, Berlin: Springer, 2013, doi: `https://doi.org/10.1007/978-3-642-36784-7`.

47. Y. Jia, "The Streaming Service Under Pandemic with the Example of Performance of Disney+," `www.atlantis-press.com`, Jan. 17, 2022, `https://www.atlantis-press.com/proceedings/sdmc-21/125968448`.

48. Statista, "eCommerce - Worldwide | Statista Market Forecast," Statista, 2023, `https://www.statista.com/outlook/dmo/ecommerce/worldwide` (accessed Sep. 24, 2023).

49. A. Galella, "Owl Labs' Survey Data on Remote and Hybrid Work Reveals about Half of Workers (51%) Say They Think More Creatively When Working Remotely versus in Office," `www.businesswire.com`, Sep. 13, 2022, `https://www.businesswire.com/news/home/20220913005266/en/Owl-Labs%E2%80%99-Survey-Data-on-Remote-and-Hybrid-Work-Reveals-About-Half-of-Workers-51-Say-They-Think-More-Creatively-When-Working-Remotely-Versus-in-Office` (accessed Sep. 24, 2023).

50. H. Wu et al., "Exploring Video Quality Assessment on User Generated Contents from Aesthetic and Technical Perspectives," arXiv (Cornell University), Nov. 2022, doi: https://doi.org/10.48550/arxiv.2211.04894.

51. S. Ahn and S. Lee, "Deep Blind Video Quality Assessment Based on Temporal Human Perception," 2018 25th IEEE International Conference on Image Processing (ICIP), Oct. 2018, doi: https://doi.org/10.1109/icip.2018.8451450.

52. Y. Zhang, C. Wang, S. Zhang, and X. Cao, "A Database for Multi-Modal Short Video Quality Assessment," ICASSP 2023 - 2023 IEEE International Conference on Acoustics, Speech and Signal Processing (ICASSP), Jun. 2023, doi: https://doi.org/10.1109/icassp49357.2023.10097274.

53. L. Mead, "How CX Leaders Win at Customer Retention," TTEC Digital, Mar. 15, 2023, https://ttecdigital.com/articles/how-cx-leaders-win-at-customer-retention (accessed Sep. 24, 2023).

54. Shahriar Akramullah, *Digital Video Concepts, Methods, and Metrics*, Apress Berkeley, CA, 2014. doi: https://doi.org/10.1007/978-1-4302-6713-3.

55. E. Court, K. Radhakrishnan, K. Ademoye, and S. Hole, "Recommendations for Big Data in Online Video Quality of Experience Assessment," *Journal of Computer and Communications*, vol. 04, no. 05, pp. 24–31, 2016, doi: https://doi.org/10.4236/jcc.2016.45004.

56. D. P. A. Menor, C. A. B. Mello, and C. Zanchettin, "Objective Video Quality Assessment Based on Neural Networks," *Procedia Computer Science*, vol. 96, pp. 1551–9, 2016, doi: https://doi.org/10.1016/j.procs.2016.08.202.

57. K. Brunnstrom, D. Hands, F. Speranza, and A. Webster, "VQeg Validation and ITU Standardization of Objective Perceptual Video Quality Metrics [Standards in a Nutshell]," *IEEE Signal Processing Magazine*, vol. 26, no. 3, pp. 96–101, May 2009, doi: https://doi.org/10.1109/msp.2009.932162.

58. A. Dowling, "VQTDL: No-Reference Algorithm for Video Quality Assessment Developed by TestDevLab," TestDevLab Blog, Jun. 21, 2022, https://www.testdevlab.com/blog/vqtdl-no-reference-algorithm-for-video-quality-assessment-developed-by-testdevlab (accessed Sep. 24, 2023).

59. A. Saha et al., "Perceptual Video Quality Assessment: The Journey Continues!" *Frontiers in Signal Processing*, vol. 3, Jun. 2023, doi: https://doi.org/10.3389/frsip.2023.1193523.

60. M. Shahid, A. Rossholm, B. Lövström, and H.-J. Zepernick, "No-reference Image and Video Quality Assessment: A Classification and Review of Recent Approaches," *EURASIP Journal on Image and Video Processing*, vol. 2014, no. 1, Aug. 2014, doi: https://doi.org/10.1186/1687-5281-2014-40.

61. C. Zhang, S. Liu, and H. Li, "Quality-guided Video Aesthetics Assessment with Social Media Context," *Journal of Visual Communication and Image Representation*, p. 102643, Sep. 2019, doi: https://doi.org/10.1016/j.jvcir.2019.102643.

62. I. Urbaniak and M. W. Wolter, "Quality Assessment of Compressed and Resized Medical Images Based on Pattern Recognition Using a Convolutional Neural Network," *Communications in Nonlinear Science and Numerical Simulation*, vol. 95, pp. 105582–105582, Apr. 2021, doi: https://doi.org/10.1016/j.cnsns.2020.105582.

63. X. Yu, Z. Tu, N. Birkbeck, Y. Wang, Balu Adsumilli, and A. C. Bovik, "Perceptual Quality Assessment of UGC Gaming Videos," arXiv (Cornell University), Mar. 2022, doi: https://doi.org/10.48550/arxiv.2204.00128.

64. D. Li, T. Jiang, and J. Ming, "Recent Advances and Challenges in Video Quality Assessment," *ZTE Communications*, vol. 17, no. 1, pp. 3–11, Nov. 2019, doi: https://doi.org/10.12142/ztecom.201901002.

65. X. Liu et al., "NTIRE 2023 Quality Assessment of Video Enhancement Challenge," arXiv (Cornell University), Jul. 2023, doi: https://doi.org/10.48550/arxiv.2307.09729.

66. W. Sun, T. Wang, X. Min, F. Yi, and G. Zhai, "Deep Learning Based Full-reference and No-reference Quality Assessment Models for Compressed UGC Videos," arXiv (Cornell University), Jun. 2021, doi: https://doi.org/10.48550/arxiv.2106.01111.

67. W. Zhou, X. Min, H. Li, and Q. Jiang, "A Brief Survey on Adaptive Video Streaming Quality Assessment," arXiv (Cornell University), Feb. 2022, doi: https://doi.org/10.48550/arxiv.2202.12987.

68. Tulsi Pawan Fowdur, M.A.N. Shaikh Abdoolla, and Lokeshwar Doobur, "Performance Analysis of Edge, Fog and Cloud Computing Paradigms for Real-Time Video Quality Assessment and Phishing Detection," *International Journal of Pervasive Computing and Communications*, Feb. 2023, doi: https://doi.org/10.1108/ijpcc-09-2022-0327.

69. T. Jo, *Machine Learning Foundations*. Springer Nature Switzerland AG 2021: Springer International Publishing, 2021, doi: https://doi.org/10.1007/978-3-030-65900-4.

70. V. Kurama, "Deep Learning on the Web with JavaScript," Paperspace Blog, Nov. 06, 2020, https://blog.paperspace.com/javascript-deep-learning-on-web-browsers/ (accessed Sep. 24, 2023).

71. D. Smilkov et al., "TensorFlow.js: Machine Learning for the Web and Beyond," arXiv (Cornell University), Feb. 2019, doi: https://doi.org/10.48550/arXiv.1901.05350.

72. M. Nagpal, "Data Preparation for Machine Learning Projects: Know It All Here," ProjectPro, Jul. 15, 2023, https://www.projectpro.io/article/data-preparation-for-machine-learning/595 (accessed Sep. 24, 2023).

73. G. Lawton, "Data Preparation in Machine Learning: 6 Key Steps," SearchBusinessAnalytics, Jan. 27, 2022, https://www.techtarget.com/searchbusinessanalytics/feature/Data-preparation-in-machine-learning-6-key-steps (accessed Sep. 24, 2023).

74. "Preparing Your Dataset for Machine Learning: 8 Basic Techniques That Make Your Data Better," AltexSoft, Mar. 19, 2021, https://www.altexsoft.com/blog/datascience/preparing-your-dataset-for-machine-learning-8-basic-techniques-that-make-your-data-better/ (accessed Sep. 24, 2023).

75. R. Sathya and A. Abraham, "Comparison of Supervised and Unsupervised Learning Algorithms for Pattern Classification," *International Journal of Advanced Research in Artificial Intelligence*, vol. 2, no. 2, 2013, doi: https://doi.org/10.14569/ijarai.2013.020206.

76. I. H. Sarker, "Machine Learning: Algorithms, Real-World Applications and Research Directions," *SN Computer Science*, vol. 2, no. 3, pp. 1–21, Mar. 2021, doi: https://doi.org/10.1007/s42979-021-00592-x.

77. L. Alzubaidi et al., "Review of Deep Learning: Concepts, CNN Architectures, Challenges, Applications, Future Directions," *Journal of Big Data*, vol. 8, no. 1, Mar. 2021, doi: https://doi.org/10.1186/s40537-021-00444-8.

78. L. Drejeris, "Mastering Neural Networks with JavaScript: A Step-by-Step Guide," Medium, Feb. 28, 2023, https://javascript.plainenglish.io/mastering-neural-networks-with-javascript-a-step-by-step-guide-f8b3dc2642ae (accessed Sep. 24, 2023).

79. S. K. Bisen, "Large Language Models (LLM): Difference between GPT-3 & BERT," Bright ML, Aug. 08, 2023, https://medium.com/bright-ml/nlp-deep-learning-models-difference-between-bert-gpt-3-f273e67597d7 (accessed Sep. 24, 2023).

80. M. Bansal, "Face Recognition using Transfer Learning and VGG16," Analytics Vidhya, Aug. 31, 2020, `https://medium.com/analytics-vidhya/face-recognition-using-transfer-learning-and-vgg16-cf4de57b9154` (accessed Sep. 24, 2023).

81. A. Sarkar, "Understanding EfficientNet — The Most Powerful CNN Architecture," MLearning.ai, May 08, 2021, `https://medium.com/mlearning-ai/understanding-efficientnet-the-most-powerful-cnn-architecture-eaeb40386fad` (accessed Sep. 24, 2023).

82. T. Feng, Z. Zhou, J. Tarun, and V. Nair, "Comparing Baseline Shapley and Integrated Gradients for Local Explanation: Some Additional Insights," `https://arxiv.org/ftp/arxiv/papers/2208/2208.06096.pdf` (accessed Sep. 24, 2023).

83. Satishsaini, "An Introduction to YOLO: You Only Look Once (Object Detection Algorithm)," Medium, Apr. 02, 2023, `https://medium.com/@satishsaini905/an-introduction-to-yolo-you-only-look-once-object-detection-algorithm-249dd82f42c` (accessed Sep. 24, 2023).

84. A. Verma, C. Kapoor, A. Sharma, and B. Mishra, "Web Application Implementation with Machine Learning," 2021 2nd International Conference on Intelligent Engineering and Management (ICIEM), Apr. 2021, doi: `https://doi.org/10.1109/iciem51511.2021.9445368`.

85. J. Huang and L. Cai, "Research on TCP/IP Network Communication Based on Node.js," AIP Conference Proceedings, 2018, doi: `https://doi.org/10.1063/1.5033779`.

86. S. Tilkov and S. Vinoski, "Node.js: Using JavaScript to Build High-Performance Network Programs," *IEEE Internet Computing*, vol. 14, no. 6, pp. 80–3, Nov. 2010, doi: `https://doi.org/10.1109/mic.2010.145`.

87. S. M. Zhao, X. L. Xia, and J. J. Le, "A Real-Time Web Application Solution Based on Node.js and WebSocket," *Advanced Materials Research*, vol. 816–17, pp. 1111–15, Sep. 2013, doi: `https://doi.org/10.4028/www.scientific.net/amr.816-817.1111`.

88. G. Jadhav and F. Gonsalves, "Role of Node.js in Modern Web Application Development," *International Research Journal of Engineering and Technology* (IRJET), vol. 07, no. 06, pp. 6145–50, Jun. 2020.

89. P. Patel, "Top 10 Node.js Use Cases by Simplior," Simplior Technologies Private Limited, May 09, 2022. `https://www.simplior.com/node-js-use-cases/` (accessed Sep. 24, 2023).

CHAPTER 2

Network Traffic Monitoring and Analysis

This chapter covers the foundations of Network Traffic Monitoring and Analysis (NTMA) by providing a thorough review of its fundamental ideas while highlighting the crucial part it plays in evaluating network activity and performance. Coupled with the massive influx of users onto our networks, activities such as web browsing and video streaming have gained increasing popularity, further necessitating measures and standards for maintaining the quality of service (QoS) of network environments. It is thus crucial to forecast network traffic parameters, such as upload and download rates, that have a direct influence on QoS. This is followed by exposure to cutting-edge NTMA approaches currently being used by worldwide organizations and major telecommunications operators. Moreover, an overview of NTMA approaches for categorizing and forecasting network traffic is also given. This chapter thus focuses on the theoretical aspects of the framework upon which the NTMA proceedings in this book are based.

2.1 NTMA Fundamentals

The internet is witnessing an exponential rise in network traffic, which refers to the volume of data transmitted through a network at a certain point in time [1]. This growth is attributed to the expanding scale and sophistication of the internet. Monitoring network performance, ensuring security, and managing resources are essential components of system administration [2]. This topic leads into the realm of NTMA, where current scientific investigations mostly focus on the application of deep learning (DL) models, including convolutional neural networks (CNNs), recurrent neural networks (RNNs), and long short-term memory (LSTM) analysis. Generally, the network

© Tulsi Pawan Fowdur, Lavesh Babooram 2024
T. P. Fowdur, L. Babooram, *Machine Learning For Network Traffic and Video Quality Analysis*,
https://doi.org/10.1007/979-8-8688-0354-3_2

development life cycle (NDLC) is used to design a cycle for the network monitoring system, which does not have a definite starting point or ending, while catering for the following areas [3]:

- Evaluate the requirement for conducting research, identify existing issues, and break down network topologies.

- Develop a monitoring strategy for a particular timescale.

- Execute the research through monitoring, analysis, and implementation.

- Capture and record the monitoring outcomes through logging.

- Control and assess the results.

- Make network management recommendations.

With the addition of analytics, fueled by the recent advances in ML, this list grew further, with the possibility of having predictive outcomes for network metrics. This causes a paradigm shift, which allows for more network management options through optimized and ML-powered resource allocation. Figure 2-1 depicts the foundational structure around which some previous studies have based their NTMA process [3].

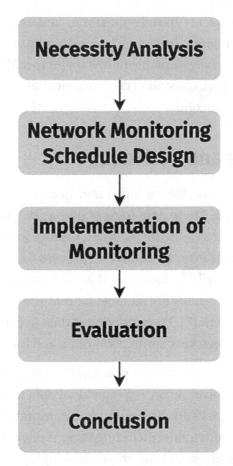

Figure 2-1. *NTMA research method*

The end of 2020 saw the digital universe being made up of a total of 44 zettabytes, distributed throughout half a million networks [4]. One of the core principles of NTMA involves conducting extensive research and using effective strategies to first examine and then quantify interactions within these communication systems. This process entails the rapid acquisition of data, subsequent adaptation, and data analysis, all while ensuring there is little impact on the integrity of the original data [5].

According to Cisco, the data-gathering method is often classified into two main categories, namely active and passive measurements, out of which 90 percent of the existing techniques are flow based [5]. This refers to the ongoing monitoring of metrics that traverse the established framework, including the frequency of packets sent and received per second, as well as the pace at which bandwidth fluctuates. Conversely, active measures are known as packet-based approaches and include the examination and extraction of packet information, including headers and payload. This approach is implemented using commercially available hardware [6] to address the disparity between the rates of deep packet inspection (DPI) and writing data to disk, which may reach several gigabits (Gb) per second.

2.1.1 Data Sources and Collection

In the digital sphere, data sources exhibit a wide range of diversity, mirroring the expansive nature of the network. These data sources play a pivotal role in offering valuable information pertaining to network behavior, performance, and security. Therefore, a robust framework is usually required to process this big influx of data. However, the highlighted downfall of existing big data technologies is their tendency to be built around offline models, usually not suitable for real-time traffic analysis, which is one of the fundamentals of NTMA [7]. The gathering of data in NTMA is often centered on three main sources, namely packet capture, flow data, and log files [8], as follows:

- **Packet Capture**: Packet capture is considered the most detailed approach since it involves recording individual data transmissions as they go through a network. The output is a comprehensive examination of network traffic, including various elements such as payload content, headers, and communication patterns. Packet-level analysis is an essential component in the examination of delicate network problems or the execution of forensic investigations. However, it may result in substantial amounts of data, requiring robust storage and processing capabilities.

- **Flow Data**: In contrast, flow data consolidates network activity into flow statistics. Flows are indicative of communication sessions that occur in a single direction, frequently structured based on shared characteristics, such as source and destination IP addresses, ports, and protocol. The process of aggregating data significantly decreases the bulk amount being processed while preserving essential information for NTMA. This is highly beneficial in the areas of traffic profiling, bandwidth monitoring, and anomaly identification. Flow data is often generated by means of protocols such as NetFlow, sFlow, and IPFIX.

- **Log Files**: Log files are an additional source of data within the context of NTMA. In computer networks, components such as network devices, servers, and applications often produce logs with the purpose of documenting noteworthy occurrences and actions. These logs provide significant value in terms of comprehending network activity and discerning security issues. Contextual information plays a crucial role in the analysis of individual events, since it enables the tracking of the sequence of events that led to a network anomaly.

The effectiveness of NTMA is contingent upon the efficient and dependable acquisition of data from the aforementioned sources. Packet capture tools such as Wireshark and tcpdump are often used by network administrators and analysts in need of instantaneous insight into individual packets [9]. Both tools are heavily used depending on features such as battery, processor, and memory usage. Flow data gathering is facilitated by protocols like NetFlow that enable effective analysis while minimizing computing load [10]. Log files, which are produced by a variety of network devices and systems, are essential for the purpose of monitoring events and resolving problems [11]. In this dynamic space, there is a growing trend among enterprises of using software solutions that provide centralized data storage and processing capabilities. The aforementioned solutions optimize the process of data collection, facilitating the ability of network professionals to equip themselves with an array of data for maintaining network integrity and safeguarding against potential security breaches.

2.1.2 Key Metrics

Service providers and network administrators often require key performance indicators (KPIs) or key quality indicators (KQIs), which at first glance represent the overall performance of the network and can also be put under the microscope further. These metrics are often housed in alerting dashboards in Network Operations Centers (NOCs) and Service Operations Centers (SOCs) and should therefore be reliable, given their responsibility as the first point of contact in detecting network issues such as slowness, saturation, and service outages. These key metrics are discussed in Table 2-1 [8, 12].

Table 2-1. *Key Metrics for NTMA*

Key Metric	Description
Bandwidth Utilization	It acts as a measure of the available network bandwidth in use at a specific point in time, as a percentage. It is thus an essential indication of how well the network resources are distributed and whether or not data transmission is impeded by congestion. Excessive bandwidth usage can indicate that optimization or capacity planning and growth are necessary.
Packet Loss	As a direct indicator of user experience, packet loss represents the quantity of data packets that have failed to reach their intended recipients. In the case of real-time applications such as web conferences or video calls, it is imperative that packet loss is low. Conversely, a high loss rate would indicate poor performance in applications such as streaming and social media applications.
Latency	On the same wavelength, latency is the amount of time that data packets take to move from the source to their destination. Applications such as video conferencing and online gaming are prime examples of services requiring low latency. Excessive delay causes these programs to be unusable.
Jitter	The variation in the timing arrival of packets is known as jitter, and may be used to indicate network congestion or packet routing abnormalities. Consistent jitter rates are essential for real-time applications such as Voice over Internet Protocol (VoIP) to preserve conversation quality and enable seamless data flow.
Throughput	As a measure of the amount of data delivered to a destination over a network in a given time period, throughput represents the network's ability to efficiently convey data. Its monitoring assists in detecting unexpected variations in data transfer speeds, which may indicate network difficulties or attacks.

It is crucial that networking professionals comprehend and monitor these vital indicators given that they allow anomalies to be quickly addressed, user experience to be improved, and network security to be strengthened [13]. This fulfills enterprises' need to diagnose issues within service-level agreements (SLAs) and maintain a secure and stable digital environment. NTMA relies on precise measurement and interpretation, together with proper alarming methods, to perform optimally and maintain the greatest degree of security.

2.1.3 Data Preprocessing and Cleaning

With a broad and dynamic digital ecosystem, network data streams are diverse and often noisy. Data preprocessing and cleaning are essential procedures in ensuring that the data obtained before analysis is correct, relevant, and free of errors. Some critical processes are explained next [14, 15]:

- **Data Collection and Aggregation**: Gathering data from several sources, including packet capture, flow data, and log files, is the initial stage in data preparation. With the aim of minimizing the bulk influx and expediting computation, collected data is frequently combined into digestible portions. Time-based sampling is one of the aggregation techniques used to combine data into relevant periods for study. In short, data is generally collected at pre-set time intervals.

- **Data Normalization**: Since network data has a wide range of formats and unit types, data normalization ensures consistency by allowing direct comparisons between measurements from various sources and devices. This opens the possibility to analyze data meaningfully.

- **Missing Data Handling**: Periodically, gaps may be present in data sources, leading to biased analysis. This is handled through methods such as data imputation, which estimates missing values using available data.

- **Noise Reduction**: Noise and abnormalities are frequently the result of equipment malfunctions or transmission errors. Through techniques like filtering and smoothing, anomalies and outliers are removed to guarantee the correct representation of network activity.

- **Data Transformation**: Scaling and logarithmic transformation techniques are some examples used to ensure that information complies with analysis methodologies' presumptions and enable insightful interpretation.

- **Data Cleaning**: This involves finding and fixing errors and discrepancies in the dataset by dealing with outliers and eliminating duplicates.

The correctness of the dataset to the actual behavior and representation of the network is one of the most crucial factors in the accuracy of NTMA and therefore, cannot be overstated.

2.1.4 Network Topology and Architecture

The fundamental basis for any study in the field of NTMA is the network's topology and design. This part explores the pivotal constituents of a network.

- **Network Topology**: *Topology* refers to the configuration of devices and interconnections between them in a network. These include star, bus, ring, and mesh configurations, which have a significant influence on data gathering, traffic flow management, and security monitoring. To effectively acquire network traffic, it is necessary to strategically position monitoring equipment. The study of traffic flow is also influenced by the topology where the transmission of data packets is affected by latency, jitter, and packet loss. Topology-aware analysis is a methodology that aims to guarantee that the performance of a network and its adherence to QoS standards are in accordance with predetermined expectations [16]. Likewise, by understanding the vulnerabilities of the network, managers are able to enhance security procedures and concentrate monitoring endeavors on points that are of utmost importance.

- **Network Architecture**: The architecture encompasses the arrangement of devices, protocols, and services that together form the network. This architectural framework can either be a conventional hierarchical design or a modern software-defined network (SDN) [17]. This choice has a direct impact on NTMA. Traditional systems sometimes need the retrieval of data from many sources, but SDN architectures often consolidate data collection, hence streamlining NTMA activities. Security issues and architecture are highly interconnected. The introduction of complex network topologies has the potential to create security vulnerabilities, but SDN systems have the capability to provide improved security controls by means of centralized policy enforcement [18]. Likewise, SDN offers dynamic and customizable functionalities that are in accordance with the changing requirements of networks. An overview of an SDN controller is shown in Figure 2-2.

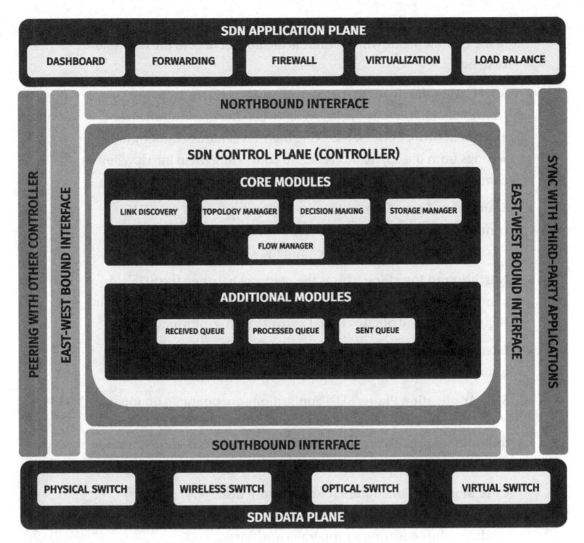

Figure 2-2. *Overview of an SDN Controller [18]*

A breakdown of the main components involved is as follows [17, 18]:

- **Data Plane:** The network packet transfer is handled by the data plane, which is made up of networking equipment, including routers, switches, and forwarding units. It is possible to have dynamic control over data forwarding in SDN since this plane is programmable and obtains signals from the controller.

- **Control Plane**: The routing choice of network traffic lies in the hands of the control plane, which controls the flow of data to the data plane switches.

- **SDN Controller**: In SDN, the controller sits at the heart of the control plane. It acts as the bridge between the application layer and the network devices, receiving high-level network policy and settings from the applications before converting them into low-level commands for the data plane equipment.

- **Southbound Interface**: This is the pathway between the SDN controller and the data plane equipment. Protocols such as OpenFlow are used to convey commands.

- **Northbound Interface**: This is the pathway between the SDN controller and the SDN applications, allowing them to convey their requirements and obtain crucial network status information.

- **East-West Bound Interface**: It enables controllers to synchronize their views and exchange information to maintain consistency.

- **SDN Application Plane**: SDN applications are operated on top of the SDN controller, acting as leverage for the global view, such that multiple services and policies can be implemented. Some examples include load balancing and traffic engineering.

- **Core Modules**: Functions such as network discovery, topology management, and routing commands are handled by the core modules, thus ensuring a proper network view.

- **Additional Modules**: Functionalities of the SDN controller can be extended to provide support to tailored use cases, such as for improving security and QoS.

- **Synchronization with Third-Party Apps**: Third-party applications allow integration with other network services and business applications, providing a malleable network management oasis.

- **Peering with Other Controllers**: To create a distributed SDN environment, controllers may be paired with one another to enable an exchange of information and the maintenance of an accurate network view.

2.1.5 Data-Driven Analytics

The use of data-driven analytics, coupled with NTMA, represents a diversion from conventional rule-based methodologies, as it emphasizes the extraction of insights directly from recorded network data. This approach uses the big data produced by devices, flows, and communication patterns, along with sophisticated analytical methods, to identify threats, abnormalities, and security risks, in turn allowing the following:

- **Data-Driven Insights**: Network experts are empowered with the ability to extract valuable insights from extensive volumes of data produced by network devices. Patterns can be discerned and anomalies can be detected with ML, statistical models, and data mining approaches. The primary benefit of using data-driven analytics lies in its capacity to forecast network activity, including traffic trends, bandwidth use, and prospective performance challenges, and its ability to take into consideration not only traffic and congestion parameters, but also the individual properties of applications, while simultaneously factoring in user behavior [19].

- **Security Threat Detection**: Through the ongoing analysis of network data, it is possible to quickly detect irregularities that may signify security breaches, attacks, or illegal access. ML models provide the capability to acquire knowledge on the typical patterns shown by network traffic, enabling them to detect and signal anomalies when variations from the norm are encountered. In some cases, Big-DAMA can speed up analysis and detection by a factor of 10 compared with the typical Apache Spark cluster [20].

- **Resource Optimization**: Companies have the ability to enhance the allocation of resources to guarantee that mission-critical applications are provided with the required bandwidth, which ultimately leads to an improvement in both user experience and operational efficiency [21].

61

- **Network Performance Enhancement**: Performance bottleneck and latency concerns are accurately detected through real-time parameter monitoring, which is followed by the implementation of appropriate remedial measures. For example, with data at the forefront of applied AI in 6G networks, descriptive, diagnostic, predictive, and prescriptive analytics are envisioned. This means having an in-depth view of channel conditions, traffic profiles, resource availability, future locations of users, network slicing decisions, and resource allocation [22].

- **Operational Efficiency**: Network engineers are equipped with the ability to make well-informed judgments, automate mundane operations, and rapidly address concerns, hence minimizing downtime and mitigating the likelihood of network disruptions.

The knowledge obtained from network data has evolved beyond mere observations, and now serves as actionable information that enables network specialists to make trained decisions, efficiently manage network resources, and ensure the security and reliability of their digital services.

2.1.6 Supervised Learning for Traffic Classification

The classification of network traffic into discrete categories is a fundamental undertaking of NTMA, as it facilitates the comprehension of the many data streams that traverse a network. This section examines the importance of supervised learning in the context of traffic classification, investigating the transformative impact of ML algorithms in the process of identifying and labeling network activity. Conventional approaches were reliant on port or protocol data that often proved to be insufficient for accurately discerning applications that use these methods dynamically. Supervised learning is a branch of ML that is trained upon labeled datasets, i.e., predetermined categories or labels, such as video streaming, file transfer, or web browsing. The classification process can be summed up through the following points:

- **Data Collection**: Capturing packet-level or flow-level information.

- **Data Labeling**: Manual categorization of traffic into distinct labels.

- **Feature Extraction**: Identification and extraction of relevant characteristics and traits.

- **Model Training**: Training various models with the labeled datasets.

- **Traffic Classification**: Using the model trained to classify unfamiliar network traffic.

The advantages of supervised learning in traffic classification are manifold. It provides the capability to identify a broad spectrum of applications, even when they employ non-standard ports or encryption. Additionally, it can adapt to emerging applications without requiring manual rule updates, making it suitable for the ever-evolving network environment. One such example is shown in [23], where an active form of ML is used to cut down the need for labeled sets of values before model training. The algorithm proposed actively chooses the instances that require labeling and goes on to achieve a high accuracy with a small dataset.

Therefore, with powerful datasets equipped with distinguishable features with regard to the labels assigned either before or actively, classification systems have the ability to adapt to developing applications without the need for human intervention, in turn allowing real-time network environments to continuously evolve. This gives rise to dynamic traffic management and the ability to respond immediately to attacks. On the same wavelength, patterns related to security concerns facilitate the timely identification of intrusions, data exfiltration, and other unauthorized activity. An example is the multi-feature fusion-based incremental technique for detecting unknown traffic, as shown in [24]. Similarly, the precise classification of network traffic enables companies to effectively manage network resources according to application priorities, guaranteeing that critical applications obtain the necessary bandwidth and prioritizing the delivery of essential services.

2.1.7 Unsupervised Learning for Anomaly Detection

Networks are characterized by their dynamic and intricate nature, whereby a substantial number of information exchanges are often observed. The presence of anomalies may serve as an indication of security breaches, network problems, or developing dangers, and it is of utmost importance to promptly identify them [25]. Unsupervised learning algorithms are used to examine network traffic in the absence of pre-established categories or labels through the following steps:

- **Data Collection**: Capturing packet-level or flow-level information.

- **Feature Extraction**: Identification and extraction of relevant characteristics and traits.

- **Model Training**: Clustering or dimensionality-reduction approaches are used to discern patterns or structures inherent in the data.

- **Anomaly Detection**: Identifying deviations from the established patterns to indicate probable abnormalities.

Unsupervised learning models provide the ability to adjust and accommodate modifications in network conditions, making them appropriate for dynamic settings characterized by the presence of emergent threats or anomalies that may not adhere to predefined trends. Finding suspicious behavior can be useful in early warning systems by analyzing even the most up-to-date datasets and streaming data. Potential application fields include intrusion detection, payment fraud detection, public safety, complex system monitoring, and medical data analysis. Lastly, with unsupervised learning, anomaly identification requires no prior training, and the data is evaluated entirely based on its inherent structure, while anomalies are frequently graded based on their degree of suspicion [26].

2.1.8 Predictive Analytics

The capacity to predict network behavior is a fundamental element of proactive network management given the rise of new frameworks, including Internet of Things (IoT), Internet of Vehicles (IoV), and 6G. Predictive analytics is usually referred to as network traffic prediction (NTP), which is a subset of NTMA and traditionally relies on time-series forecasting (TSF), i.e., the creation of models capable of estimating future traffic volumes through the identification of existing dependencies between past and future data [27]. The advantage of TSF methodologies is their low complexity, but the influx of network traffic has also called for non-TSF frameworks, which act upon flow and packet header data, to forecast future flows, rather than network parameters. There is thus a back-and-forth shift between forecasting flows and network traffic, depending on the use case. Some popular data collection and models used to create trained models for NTP are as shown in Figure 2-3.

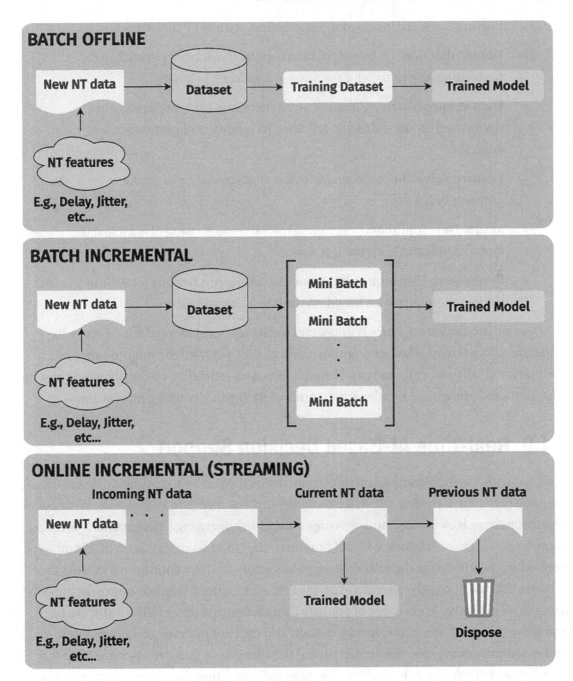

Figure 2-3. Existing frameworks for NTP [27]

The following steps are involved in the creation of an NTP framework:

- **Data Collection**: Gathering data in time-series format pertaining to performance, events, and resource utilization.

- **Data Preprocessing**: Cleaning, normalization, and preparation are performed on the dataset to enhance its quality and pertinence for analysis.

- **Feature Selection**: Selecting relevant characteristics to depict network behavior.

- **Model Development**: Developing the ML model using historical data and a concept to keep track of time.

- **Prediction**: Using the trained model to make predictions for future trends; these are often based on the latest available data.

Overall, predictive analytics is important for service providers when it comes to proactive allocation of resources, optimization of the QoS, and the maintenance of optimal network performance [28]. Through forecasted insights, key applications can be allocated the appropriate bandwidth in terms of their priority usage by customers.

2.1.9 Real-time AI-Based Decision Support

Coupled with the increasing number of interconnected devices, worldwide networks are evolving at an alarming rate in terms of novelty, volume, and complexity. With novel applications such as autonomous sensing, computing, learning, analytics, and perhaps commercial decision-making without human oversight, the combination of AI and networking is critical for digitalization, sophistication, and the continuous evolution of services [2]. Traditionally, telecom experts build and manage wireless networks, heavily using their extensive knowledge of network design, consumer mobility, traffic patterns, and wave transmission frameworks to design and execute policies and procedures that essentially guarantee the uninterrupted functioning of massive systems. With the introduction of smaller cells and cutting-edge radio technologies, network structures have become much more complicated. In addition to being more challenging to forecast, traffic patterns and radio transmission models are becoming more challenging to assess in light of newly available spectrum frequencies and highly populated networks. Therefore, AI is in a dominant position as it comes to help operators facilitate the rollout and upkeep of wireless and mobile networks [29]. Furthermore, AI has been given the confidence for autonomous decision-making and analysis capabilities, and is regarded as vital in ensuring zero-touch provisioning of these complex networks.

A strategy focused on numbers will gradually take over as a consequence of the incorporation of advanced intelligence into applications, networks, and organizational structures. This approach will increase automation, dependability, and efficiency to an unprecedented degree. Telecom companies will have more control and be able to use analytics on collected data to more effectively manage their networking infrastructure. Each service may be made more safe, dependable, and robust by customizing and ultimately mapping certain functions to network behavioral patterns [30]. This will further propel the evolution of both Wi-Fi and mobile services in terms of both societal and industrial improvements. Multiple sites and phases of decision-making processes occur inside a massively distributed architecture. While certain actions involve intense and well-planned judgments that change the behavior of the whole network and are thus driven by the gathering of data from several locations, others are determined by strict-control mechanisms with minimal latency and that rely on local data [31]. These choices frequently fall into a crucial category, necessitating prompt attention. The dispersed nature of the topology should serve as the foundation for the governing intelligence so as to continually and precisely feed this massive, universal structure.

In the case of IoT sensors used for autonomous vehicles, industry 4.0 and manufacturing operations, smart cities, and health care, data produced near the edge, or from the device's interfaces, often needs instantaneous processing. There may be situations where it is not the greatest idea to move this data to a centralized cloud because of policies that restrict where data may be kept. This suggests that data might not even be sent in some circumstances. These are the exact cases where real-time NTMA needs to be sharp, straightforward, reliable, and lightweight, while being housed at the local site itself, without needing a client–server paradigm. Learning about global data patterns directly from numerous networking nodes, without transferring them to a centralized system, is one potential solution to this challenge. It will also soon be necessary to find appropriate methods for sharing newly learned information while guaranteeing prompt transmission across devices. The collaboration of AI algorithms, ML, and other foundations in the wireless ecosystem can unlock the door to breakthroughs, including self-management, self-optimization, and self-evolution.

2.2 Existing NTMA Applications

In today's fast-paced digital landscape, enterprises primarily depend on NTMA solutions to effectively uphold the integrity and optimal functioning of their network infrastructures. Network management and monitoring tools are critical to the success of businesses, helping IT teams reduce downtime and identify bottlenecks, thus boosting the efficiency of troubleshooting efforts, in addition to finding KPIs to improve their QoS and QoE. With the focus being not just on monitoring, security professionals and administrators aim to make data-driven decisions, establish resilient security measures, and provision uninterrupted digital experiences to end users. The following is a brief description of existing software commonly used by corporate companies for NTMA around the world [32].

2.2.1 SolarWinds NetFlow Traffic Analyzer

As a prominent entity in the sphere of network surveillance, SolarWinds [33] offers a comprehensive NTMA package that encompasses the automatic compilation of extensively customizable and detailed reports. Efficient sorting characteristics, such as evaluating the proportion of bandwidth used by applications over a certain duration, can be implemented to list and categorize network-based applications. In a nutshell, real-time traffic data is collected from ongoing streams and transformed into visual representations, such as tables and charts, to depict the operational dynamics of the business infrastructure. The software offers various versions that include functions such as traffic analysis and bandwidth monitoring. There is also support for VMware vSphere distributed switch, and the capability to generate warnings based on specific types of network traffic. It is usually used for the purpose of gathering data flow and identifying sophisticated applications [34]. However, the incorporation of features such as traffic forecast and classification pertaining to network consumption is absent. The framework concentrates mainly on observational attributes, including the correlation of cross-stack IT data and the comparison of data in a side-by-side manner.

2.2.2 Paessler PRTG Network Monitor

With the aim of making the life of IT professionals easier, Paessler Router Traffic Grapher (PRTG) [35] provides a descriptive and informative overview of networks, systems, and applications. With Android and iOS applications for on-the-go monitoring, alerts can be received and handled easily on one's mobile phone. The distinguishable features include customizable dashboards that the client can create as well as select from a variety of templates that suit the need of the moment, such as presentations with the management team around the office. This ranges from creating fully operational dashboards with integrated and dynamic maps, to adding custom HTML elements. These features can be joined together to create an extensive, real-time network management tool that caters for different locations, thus providing an eagle-eye view of the area. Remote site monitoring and bandwidth analysis functionalities are also present to identify bottleneck sources. The package also offers automated browsing tests, which can be performed on any website to yield parameters such as load times, response times, and whether the site is up or down. On the same wavelength, the status of Domain Name System (DNS) servers can be checked. An example is shown in Figure 2-4, where the HTTP loading times of Facebook are measured and plotted. With cloud support and partnership with hundreds of companies, such as IBM Cloud and Citrix Systems, auto-discovery mode allows configurations to be made in an instant, with pre-configured sensors. Moreover, protocols such as ICMP, SNMP, WMI, and HTTPS are used for the data collection process. This can be applied to several scenarios, such as health-care systems and smart cities [36]. As a standalone application, PRTG does not implement ML algorithms. With the addition of pre-built modules using the Crosser platform, which is one of the company's alliances, analytics can be integrated into the software.

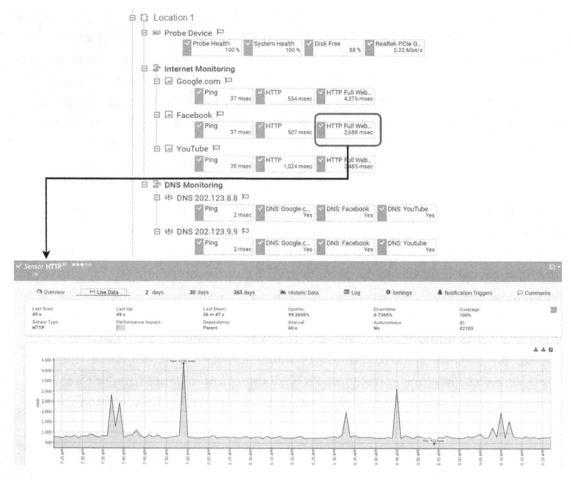

Figure 2-4. *Example of Paessler PRTG network monitor for monitoring Facebook HTTP loading times*

2.2.3 Wireshark

Wireshark [37] is widely recognized as a renowned open-source network packet analyzer tool. This means that it allows the inspection of network packet data. Although not a full-fledged NTMA tool, it significantly contributed to network traffic research as a stepping stone, allowing packet information to be captured in a timely manner, and is essential in resolving network-related problems, together with the examination of traffic patterns. A wide variety of network protocols is supported, with the in-built capability to interpret and present packet contents in a format easily understood by humans. Some examples include IP, TCP, HTTP, and SNMP [38]. This aids in managing network data flows and discovering blockages and problems. Real-time packet capture is thus the highlight of Wireshark, with subsequent analysis requiring a third-party plugin or an entire framework. Wireshark hence only covers the monitoring aspect and is a great resource for network managers, security specialists, and students who engage in the intricate aspects of NTMA, since it is open source. An example of UDP packet filtering through Wireshark is shown in Figure 2-5.

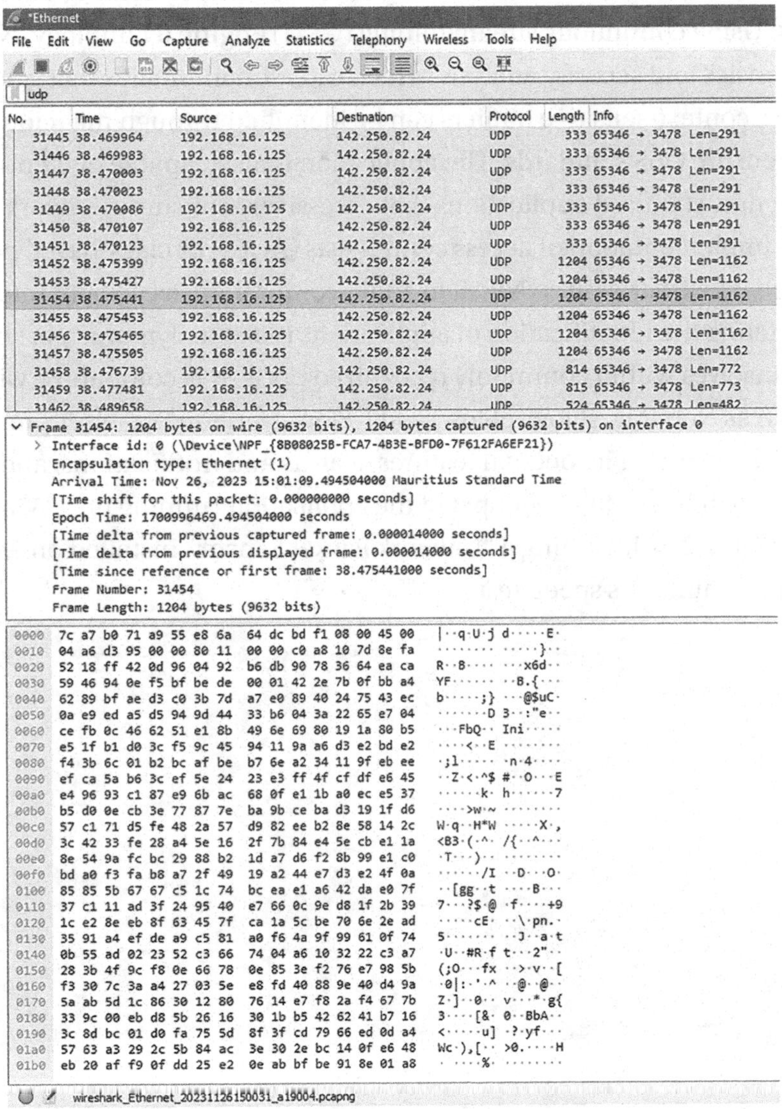

Figure 2-5. *Example of Wireshark UDP packet filtering*

2.2.4 ManageEngine NetFlow Analyzer

NetFlow Analyzer [39] is a full traffic analysis tool that focuses on flow technologies to demonstrate how network bandwidth fares in real time, with the primary aim of optimizing networks through capacity planning and security monitoring. It helps gather, analyze, and generate reports on bandwidth utilization and serves as a reliable collaborator in increasing the efficiency of global networks by drilling down to interface-level details. Using continuous stream mining (CSM) engine technology, NetFlow allows network forensics and security analysis, which cater for anomalies that get past firewalls. Furthermore, context-sensitive outliers can be identified through malicious network behavior to ensure QoS standards. The application is also capable of identifying and categorizing non-standard applications that excessively consume network bandwidth through the implementation of access control lists (ACLs) or class-based policies. This can be paired up with the Cisco NBAR to attain comprehensive insight into Layer 7 traffic, facilitating the identification of applications that use dynamic port numbers or conceal themselves within commonly recognized ports. It is compatible with major flow formats, such as NetFlow, sFlow, cFlow, and J-Flow. NetFlow Analyzer provides IP SLA monitoring, deep packet inspection features, alarms and notifications, and distributed monitoring capabilities [40]. A snippet of the application running on a Windows device is shown in Figure 2-6, including a "Network Packet Sensor" installed on another device that performs continuous speed tests.

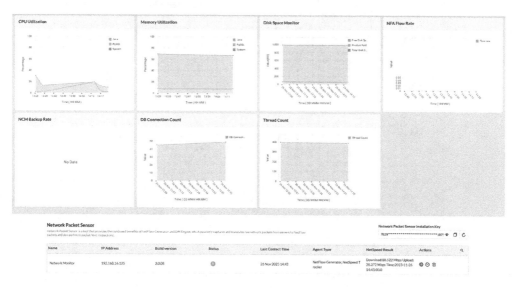

Figure 2-6. *Example of NetFlow Analyzer system performance dashboard and network packet sensor*

2.2.5 Site24x7 Network Monitoring

Site24x7 Network Monitoring [41] is a cloud-based network monitoring platform that encompasses IT systems, programs, and user activity. The network monitoring part of the utility can automatically identify all networked devices. Site24x7 creates an equipment inventory based on the search results and subsequently dynamically generates network topology maps. When new equipment is introduced, relocated, or discarded, the list of items and maps is instantaneously updated. The software goes beyond simple connection and availability checks. It also verifies metrics concerning connectivity, such as downtime duration, last downtime, response times, and download times of websites. A typical scenario is shown in Figure 2-7, where these metrics are measured in real time for a user-input web page.

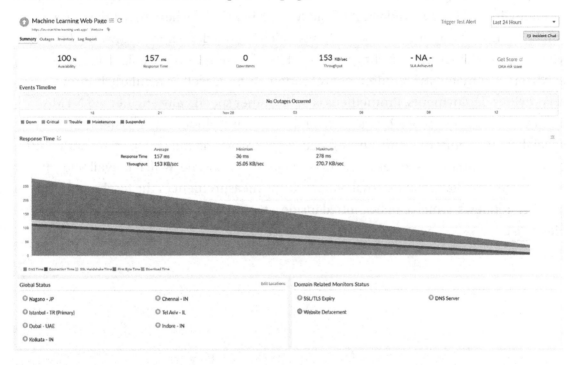

Figure 2-7. *Example of Site24x7 web page monitoring*

Users have the ability to choose the specific services they want to monitor and to establish performance criteria that will activate alert notifications. The auto-discovery functionality of Site24x7 is facilitated by the use of the Simple Network Management Protocol (SNMP). To get real-time information from network devices that are configured to route data through SNMP, Site24x7 acts as the imperative SNMP manager. The Simple

Network Management Protocol (SNMP) has a mechanism whereby device agents are able to transmit notifications to the manager upon detecting a severe situation on the monitored device. The communication is transformed into a warning or an alert within the Site24x7 dashboard, contingent upon the degree of seriousness assigned to the message by the agent. The Site24x7 plans exhibit a notable emphasis on the surveillance of web applications, effectively integrating these services with network and connectivity assessment capabilities. Thus, enterprises that indulge in web development and operations largely employ Site24x7 as their network management tool.

2.2.6 Prometheus

On top of providing the same features as mentioned for the previous software, Prometheus [42] is equipped with interactive web interfaces where network interfaces can be individually monitored, with alarms following the detection of unwanted traffic set by the user. Since the software is open source, the community is able to create exporters that match the requirements of certain users as well as regularly improve on previous deployments. Prometheus is unlike other specifically engineered NTMA software in the sense that it is designed for dynamic and cloud-native systems. It specializes in monitoring and alerting features by extracting metrics from set endpoints and storing them in a time-series database. One such example, which is available without the installation of external plugins, is the measurement of the total number of HTTP requests, which can be plotted inside a typical web browser, as shown in Figure 2-8.

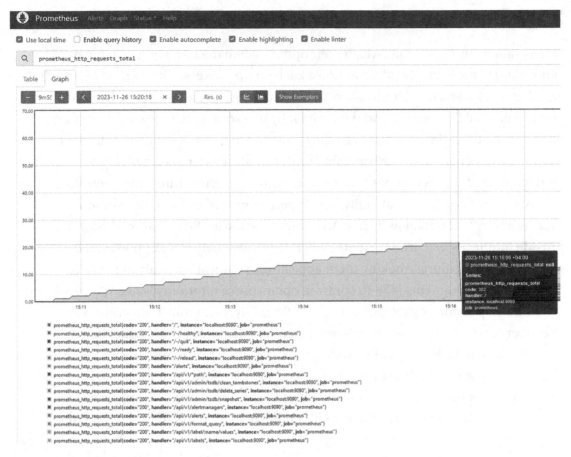

Figure 2-8. *Example of Prometheus graphing tool for monitoring HTTP requests*

With a robust query language, it facilitates the acquisition of profound insights into the accumulated data, making it a suitable option for entities that use microservices and containerized applications. Prometheus has an elaborate ecosystem that encompasses exporters designed for a range of services and components. This feature greatly simplifies the process of integrating the tool into pre-existing settings, ensuring a smooth and efficient transition. Additionally, it provides connection with Grafana, allowing users to create personalized dashboards for the purpose of viewing the metrics that have been gathered. This technology enables DevOps teams, system administrators, and programmers to acquire a thorough comprehension of their system's behavior, promptly address problems, and maintain a high level of service dependability. Some of Prometheus's users are Docker, Soundcloud, Showmax, and Amadeus. Like the aforementioned products, traffic prediction services need to be added as tailor-made modules like Locust.

2.2.7 Commercial vs. Open-Source Solutions

Organizations must carefully weigh their options when deciding between open-source and commercial NTMA solutions. Each group possesses its own set of benefits and drawbacks, and the decision should be made in accordance with individual needs, financial limitations, and the availability of resources. Commercial solutions often incur a financial cost, although they provide extensive assistance, abundant features, and dependable performance. Typically, these systems are highly compatible with major companies that possess intricate network architectures that necessitate extensive functionalities, scalability, and expert support. Commercial solutions such as SolarWinds NetFlow Traffic Analyzer and Paessler PRTG Network Monitor provide interfaces that are easy to use, offer dedicated customer support, and have comprehensive features, making them appealing to enterprises with substantial monitoring requirements [43]. In contrast, open-source alternatives like Wireshark and Prometheus provide advantages in terms of flexibility, customization, and cost-effectiveness due to their software's being free. Smaller enterprises, educational institutions, and technologically proficient individuals who possess the necessary skills to set up and sustain these technologies have a preference for them. Open-source solutions foster the creation of a collaborative community, empowering users to actively participate in their advancement and tailor them to meet particular requirements.

2.2.8 Challenges and Considerations

The implementation of NTMA solutions, regardless of whether they are commercial or open source, can hit a range of setbacks, spanning from data gathering to data privacy. Some of these difficulties are elaborated as follows [44, 45, 46]:

- **Data Volume:** The use of NTMA tools generates significant quantities of data, analogous to big data cubes, necessitating strong storage and processing capabilities. Organizations must give due consideration to the formulation of data retention regulations and the implementation of data minimization methods.

- **Encrypted Traffic:** With the increasing use of encrypted traffic, namely HTTPS, businesses face the multifaceted task of finding a delicate equilibrium between security imperatives and privacy concerns.

- **Network Scalability:** The ability to expand and adapt continuously is a crucial consideration for businesses looking to effectively address monitoring requirements. This becomes paramount when the network undergoes expansion or encounters heightened levels of load.

- **Privacy and Compliance:** Organizations are required to conform to legislation and establish protocols aimed at safeguarding user information, while also adhering to data protection laws.

- **Integration with Existing Infrastructure:** Efficient monitoring ultimately means compatibility with various devices, protocols, and services, together with existing and new network infrastructures.

- **Alerting and Reporting:** The successful implementation of NTMA entails proper warning systems and the generation of informative reports. It is important for businesses to decide upon appropriate thresholds for alerts, and ensure that their reporting processes provide timely and relevant information.

- **Resource Allocation:** In terms of CPU and memory utilization, systems performing real-time NTMA need to provide optimal performance to yield the best results and offer reliable performance.

- **Training and Expertise:** It is recommended that organizations allocate resources for training opportunities for their personnel.

The main challenge, however, lies in the erratic nature of network traffic, which has allowed it to narrowly escape the clutches of highly accurate predictive and classification algorithms when it comes to ML. This book addresses this predicament in areas that have been rather sluggish. Through predictive features, features such as the anticipation of network-related problems are unlocked, allowing certain behaviors to be discerned. Thus, in recent years, the spotlight has been on proactive insights into network activity, forecasting abnormalities, and automating the resolution of detected issues.

2.3 State-of-the-Art Review of NTMA

According to recent data from January 2022, the worldwide population of internet users reached its highest point at 4.95 billion individuals, accounting for around 62.5 percent of the entire global population [47]. This data suggests a substantial surge, with around 192 million members added within a span of one year. The COVID-19 pandemic has prompted a significant shift among enterprises toward using web-based applications and services. This transition has introduced additional complexities to effectively managing difficulties related to quality of service (QoS), quality of experience (QoE), scalability, and performance [48]. Based on Cisco's findings, it is apparent that IP video traffic constituted a significant proportion of IP traffic in 2021, accounting for around 82 percent [49]. It is anticipated that this remarkable growth trajectory will persist in the foreseeable future.

2.3.1 Background of NTMA

The concept of network traffic monitoring and analysis (NTMA) pertains to the examination of metrics related to the flow of data messages in order to detect meaningful and exploitable trends in behavior, thus enabling the real-time assessment of a network's state and functionality. For context, there is a notable 53 percent rise in the probability of individuals' accessing a web application if its loading durations do not exceed two seconds, which is even more renowned in browser extensions, which provide an even more expedient and convenient means of access requiring just a few clicks [50]. Therefore, tracking and evaluating network traffic metrics is an important responsibility for service providers and network managers. This practice is essential to effectively manage bandwidth needs and mitigate the risks posed by cyber-attacks.

2.3.2 The Rise of Machine Learning

Recent studies have revolved around the use of ML-based NTMA methods, especially in strategies that focus on port-based and load-based methodologies where performance is gauged through accuracy evaluation. The collaboration of ML and NTMA brings about a systematic examination of network traffic that makes informed predictions about patterns, characteristics, and parameters. This is crucial when assessing the QoS of networks and evaluating their levels of accuracy and precision, alongside rates of false positives and false negatives. Both prediction and accuracy detection have become easier with the advent of machine learning [51]. Correspondingly, the surge in use and flexibility of cloud platforms have vastly magnified storage and computation needs. According to Medium [52], a significant majority of global enterprises, approximately 81 percent, have implemented a multi-cloud strategy. In this context, advanced analytics techniques like the decision tree (DT) and multilayer perceptron (MLP) algorithms can be integrated into cloud computing platforms. By leveraging these platforms, real-time network data can be collected and analyzed, enabling the generation of informative and predictive outcomes. This eliminates the need for local processing power. The incorporation of ML into analytics has played a crucial role in the categorization of data into either supervised or unsupervised learning sets.

Currently, with regard to analytics, network traffic can be both classified and predicted. The former can be associated with different levels of identifying the nature of network traffic, ranging from the type of activity being performed by its users, to more advanced and computationally intensive classifications, such as identifying the exact application being used. However, network parameters such as bandwidth, latency, jitter, and website response time can be put through regression algorithms, which aim to forecast these variables at a given time [2]. A state-of-the-art review concerning related works performed by researchers in both realms is presented next.

2.3.3 Machine Learning Algorithms to Classify Network Traffic

In recent years, several researchers have been actively engaged in using ML algorithms for NTMA with the aim of identifying methodologies that provide optimal accuracy. To classify network traffic in [53], Pradhan aimed to enhance the QoS by using a number of flow characteristics as input variables. The data-preparation phase was performed using a Perl script, together with Wireshark, to generate a set of thirteen properties. The labels assigned for classification were FTP, WWW, and P2P. The trained models were artificial neural network (ANN), SVM, MLP, and sequential minimal optimization (SMO) and the experiment was performed using the Weka framework, as shown in Figure 2-9. The extracted features include the packet arrival time, packet length, protocol, and destination port number, among others. With the simulation model taking as input values in the comma-separated values (CSV) file format, it was found that ANN yielded an accuracy of 85.443 percent while SVM had a precision of 91.1392 percent, where both algorithms were subjected to different training percentages. The investigation concluded that SVM performs much better than ANN when classifying network traffic.

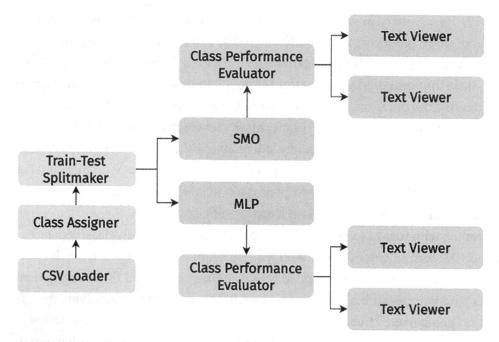

Figure 2-9. *Workflow of the proposed SVM and ANN model by Pradhan [53]*

In Shafiq et al., they analyzed the performance of a series of ML techniques aimed toward the classification of the types of applications surrounding a network [54]. The research provides a comparative analysis between SVM, C4.5 decision tree, naïve Bayes, and Bayes Net, where network traffic was first captured using Wireshark, for an interval of one minute. The applications investigated were DNS, WWW, FTP, P2P, and Telnet, together with a total of twenty-three extracted features, using the Netmate tool. The best performance from the experiment was from the C4.5 algorithm, which yielded a classification accuracy of 79 percent as well as the least training time.

In Dixit et al., an analysis [55] was conducted on the prevalent forms of internet traffic, which primarily comprised the K-nearest neighbors (KNN) and naïve Bayes algorithms, coupled with statistical network metrics like inter-packet arrival time, time to live (TTL), and packet count. The research focused on safeguarding the confidentiality of users and obtained a precision rate of 85.3 percent using a train–test split ratio of 70:30 with KNN, while naïve Bayes achieved 54.2 percent. The highest accuracy was obtained with eight neighbors for KNN. Its use is promoted for categorizing network properties based on criteria that possess comparable characteristics.

In Guezzaz et al., an analysis [56] is performed of an MLP classifier in the context of network monitoring using a three-layer architecture. They outlined a training methodology that is very efficient in determining optimal weights, with a detection technique that serves to validate the model. The performance of most classifiers is adversely impacted by the presence of huge datasets, influencing the efficiency of the sorting procedure. MLP, however, demonstrates its capability as a perceptual analysis layer that may be used in network monitoring systems to enhance the trustworthiness of data transmission. Similarly, in Abassi et al. [8], deep learning (DL) techniques are used for building network traffic management models. The study included algorithms such as MLP, CNNs, and RNNs designed for time-series data. With DL approaches historically focused on exploiting port-based, packet-based, and flow-based metrics, the authors presented and described newer traffic classification techniques.

By considering the privacy of corporate data, Bayat et al. [57] merged an array of deep learning techniques to create an intricate classification model that analyzes packet flow and payload information, as well as the time difference between packet arrivals. With the goal of identifying applications without inspecting the contents of encrypted packets, the procedure was geared toward solving the server-name indication (SNI) classification problem during transport layer security (TLS) handshakes. The procedure started by first collecting some 500,000 HTTP requests to different websites followed by storing packet capture files through Wireshark and the SSL filter property. Features such as accumulated bytes and arrival times were extracted to prepare the training dataset. The main algorithms used included CNN and gated recurrent units (GRU), which provide an extension to feed-forward neural networks by allowing the length of the sequences to be varied. A hidden state was designed to activate after each previous state, which gave rise to the adaptive characteristic of the algorithm. For Google Chrome data, the experiment resulted in an 82.3 percent accuracy using the combination CNN-RNN model. The architecture used is shown in Figure 2-10.

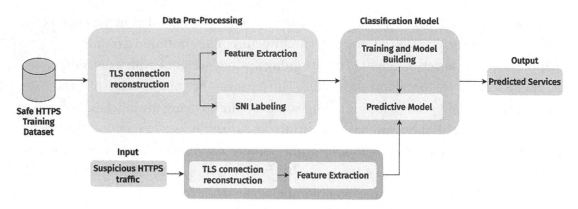

Figure 2-10. *Workflow of the proposed HTTPS identification framework by Bayat et al. [57]*

To provide a rapid and automated multimedia traffic management framework, Wu et al. proposed a CNN model that operates by self-improving through progressive learning [58]. The experiment was aimed at the immediate classification of network traffic with respect to the QoS through deep learning. Equipped with the sliding window model, parameters were extracted based on data flows, after which a 3D representation of each flow was drawn out. The CNN model was then tailored to match the variations of the selected features. In addition, techniques such as knowledge distillation and bias correction were added to the architecture to support the self-learning characteristic without having to stop the model training process to accommodate new patterns. This resulted in an increase in the classification speed compared to previous research, along with a 50 percent decrease in memory usage despite the addition of new categories of network traffic.

Regarding the use of DL for network traffic classification, in [59], the authors developed the naïve Bayesian and C4.5 decision tree methods. The paper emphasized the importance of using ML in the case of streaming, gaming, and P2P applications, where the port-based protocol identification approach cannot be relied on. Adding to the list of intricacies is the inability of traditional methods to work with encrypted traffic, which is why research turns toward the use of either statistical or ML methods. The experiment achieved a classification accuracy of about 95 percent after the implementation of a feature selection step, paired with a multilayer neural network. The algorithms were run on ten distinct categories of data, each derived from an extensive TCP bidirectional stream consisting of 249 network traffic parameters. These were categorized into eight broad traffic types for analysis given the limited number of samples available for certain applications, such as games. The DL framework used in this study, i.e., Google's TensorFlow, attained a stable accuracy rate of 98.5 percent in sample sizes ranging from 0 to 10,000, as compared with naive Bayes and random forest, which both fall short in terms of accuracy and only perform well for small sample sizes. Stability and reliability are therefore exhibited by the DL algorithm.

Labayen et al. introduced a hybrid system designed to categorize network traffic into different network states, such as online browsing, streaming, and idle behavior, by analyzing traffic trends [60]. Data is first captured using Tshark into pcap and CSV formats, before being pre-labeled into different classes, including p2p, video, web, bulk, chat, email, and voip. The suggested system, depicted in Figure 2-11, takes as inputs a continuous flow of packets containing fields extracted from the IP and TCP/UDP headers, before preprocessing them into a suitable collection of features. A three-layer hierarchical window data structure is formed in which the data is divided into temporal windows. This removes any dependency of the features on one another. Next, a three-tier classification procedure is executed to obtain the user activity linked with the classification window produced. The windows used are termed as classification, flow, and behavior. The study yielded a global accuracy of 97 percent using the hierarchical multilayer structure with the best performance obtained with the K-means algorithm embedded in the first two layers, and random forest in the last layer.

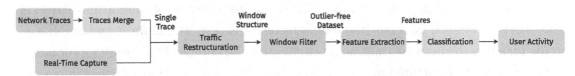

Figure 2-11. *Workflow of the proposed user activity classification framework by Labayen et al. [60]*

Bakker et al. presented a series of classifiers aimed toward the detection of distributed denial of service (DDoS) attacks [61]. Openly accessible datasets were used to construct a universally accepted benchmark for evaluating the performance of individual classifiers, specifically in terms of precision and detection rate. The experiment was implemented in a live SDN where data capture was performed on packets and bytes. Among the algorithms were random forest, KNN, quadratic discriminant analysis (QDA), naïve Bayes, and support vector machine (SVM). These were subject to comparison in both the reference standard and live network. The former saw the highest accuracy from the random forest algorithm, with SVM having the poorest performance. Meanwhile, in a live network setting, for both passive and active modes, SVM exhibited the highest accuracy, making it difficult to make a final conclusive statement.

2.3.4 Machine Learning Algorithms to Predict Network Traffic

The study by Adekitan et al. [62] revolves around the exploitation of logged internet traffic data of the Covenant University in Nigeria throughout one year. The parameters considered were the download and upload internet volume banks. Forecasting models were built around the prediction of the values for a given day, with the input dataset being the data amassed during the previous days. Four ML algorithms—namely, the tree ensemble, random forest, decision tree, and naïve Bayes—were paired up with the Konstanz Information Miner (KNIME) data mining software to generate predictive models. The researchers came forward with a precision accuracy of 55.66 percent for both download and upload IP traffic where the tree ensemble method came out on top.

Iqbal et al. presented a comparative analysis of prediction algorithms, focusing on accuracy, cost, and complexity as the key evaluation criteria [63]. The analysis of network traffic in real time was conducted using time-series approaches, artificial neural networks (ANNs), and forecasts based on wavelet transform. The authors laid emphasis on the idea that the network trace largely influences the choice of the predictor model, and so they proposed a novel metric that combines accuracy and power usage into a single variable. They also devised a score that combines performance and energy usage, and highlighted the general predictability of network traffic. It was found that double exponential smoothing (DES) and autoregression moving average (ARMA) exhibit higher prediction accuracy along with the least energy consumption with regard to online applications that require more computational resources. However, ANN seemed to be more consistent with different network traces, while the implementation of exponential smoothing (ES) as well as triple exponential smoothing (TES) only raised the intricacy without enhancing accuracy.

Similarly, Zhao and He made comparisons between long short-term memory (LSTM), gate recurrent unit (GRU), temporal convolutional network (TCN), and NTAM-LSTM [64]. The latter refers to the authors' suggested model for the prediction of network traffic, with the initial prediction being made using LSTM and an attention mechanism inserted for improved accuracy. This allowed the values of the attributes of a variety of traffic types to be calculated before the value of the weighted sum. With a train–test ratio of 90:10 on a dataset involving 10,000 instances of network metrics such as the total length of each packet in bytes, the algorithms were evaluated in terms of mean absolute error (MAE). The proposed model not only solved the incompetence of the traditional LSTM model whereby it always sets up the same length for different property values, but also performed better in terms of accuracy and prediction time. This allows the model to be placed in scenarios requiring quick network resource allocation. In short, when operating on its own, the LSTM method yielded the poorest MAE, but when paired up with the attention mechanism, it resulted in both the least performance error as well as the quickest time taken to predict.

To predict the amount of data per cell in a real-world LTE environment, Gijón et al. [65] presented an in-depth comparison between the effectiveness of supervised learning (SL) models and time-series analytical methodologies. By first collecting data from a massive dataset spanning over thirty months, the studied algorithms include random forest, neural networks, SVM, seasonal auto-regressive integrated moving average (SARIMA), and additive Holt-Winters. With the arrival of 5G projected to trigger a massive increase in congestion, the accuracy and processing complexity of the relationship between long-term data traffic and short-series forecasting was studied. Long-term records were then used to make predictions about mobile traffic, leading to the conclusion that SL approaches are superior to conventional time-series processes in terms of both precision and storage requirements.

Sun suggested an innovative approach for a smart forecasting traffic analysis tool built around a local wireless area network [66]. The research relied on the variable coefficient regression model, complemented by kernel and nearest neighbor estimation weight functions. The author labeled the algorithm as a KNN nonparametric regression prediction algorithm, summarized in Figure 2-12. Moreover, an intricate log of past records was meticulously crafted through a methodical approach to discern and identify resemblances between existing historical data and recently acquired data. The traffic was first identified to be in the region of either busy or unbusy, where the former represents erratic variations due to different usage scenarios and the latter relates to a rather stationary behavioral pattern. The algorithm demonstrated superior accuracy compared to prior investigations conducted within the same experimental setting without exception.

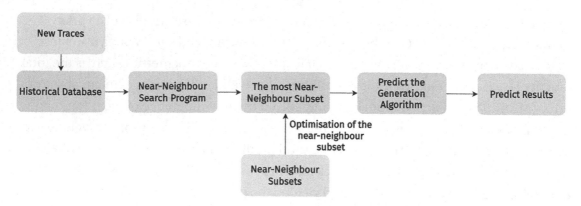

Figure 2-12. *Workflow of the proposed KNN nonparametric regression prediction algorithm by Sun [66]*

Joshi and Hadi delved into a comprehensive exploration of diverse methodologies, encompassing the realms of neural networks and data mining techniques, with the focal point centered on the suggestion of various linear and non-linear methods for forecasting [67]. The research also involved combining said methodologies to enhance prediction efficiency. Prominent datasets like DARPA and NSL-KDD facilitated the exploration of the various evaluation metrics investigated, including the autoregressive integrated moving average (ARIMA) and hybrid models. Furthermore, the review presented profoundly enriching research that explores the complex nooks and crannies of NTMA.

In a study conducted by Dalmazo et al., the researchers studied the effectiveness of predictor models in dynamic settings [68]. Specifically, they focused on the unpredictable and fluctuating traffic patterns commonly observed in modern networks. In the context of cloud computing, multiple variables were considered when evaluating a method, including efficiency, accuracy, computation time, and complexity. Six ML algorithms were tested on their ability to forecast time-series data using a static and a dynamic set of inputs. These include the simple moving average (SMA), weighted moving average (WMA), exponential moving average (EMA), Poisson moving average (PMA), autoregressive moving average (ARMA), and autoregressive integrated moving average (ARIMA) algorithms. This framework is shown in Figure 2-13. The study revealed that the implementation of a dynamic window size algorithm (DyWiSA) resulted in a significant enhancement in the accuracy of all six algorithms when compared to the use of the static window approach. The paper also recommended NTMA as a way to mitigate the impact of sudden fluctuations in historical patterns over time.

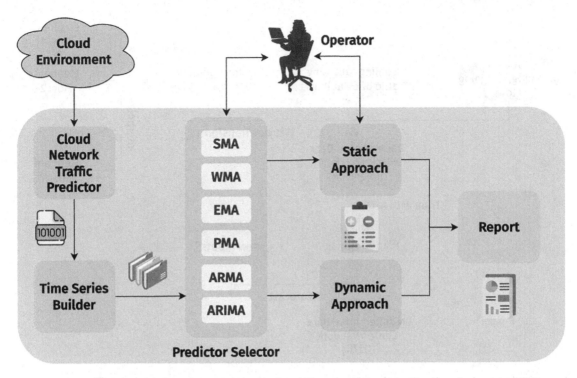

Figure 2-13. *Workflow of the proposed traffic patterns prediction framework by Dalmazo et al. [68]*

Yi Li et al. proposed the combination of wavelet transform with ANN to target an improvement in the accuracy of data transfers in inter-data-centers (inter-DC) [69]. They emphasize the importance of traffic prediction to mitigate the utilization of and bottlenecking between data centers. Time domain–captured traffic was first decomposed before being doped with explicit elephant flows and sub-link information, as outlined in Figure 2-14. This step proved to be crucial to the success of the experiment, increasing the prediction accuracy for periods of up to five minutes. Interpolation mechanisms are also mentioned for reducing the measurement costs. The main conclusion, however, is the improved prediction method that can help Baidu bring about an improvement of 9 percent on its bandwidth cap. The authors lay particular stress on the importance of continuously lowering the error through research, as a small enhancement proves to be significant to companies that make regular financial decisions.

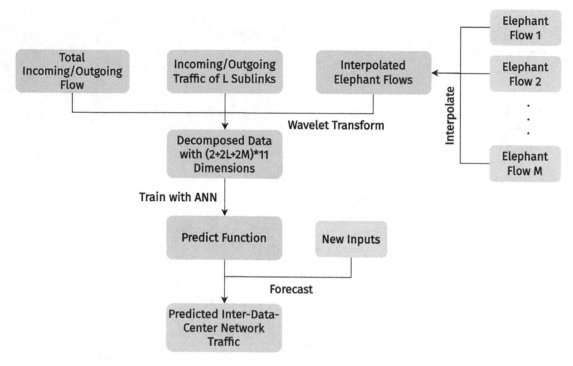

Figure 2-14. *Workflow of the proposed inter-data-center network traffic prediction model by Yi Li et al. [69]*

A number of time-series prediction algorithms were introduced by Gowrishankar et al. for predicting wireless network traffic in [70]. The application of statistical frameworks and neural networks has been a focal point in the development of procedures for estimation across various timestamps. The analysis also encompassed the evaluation of the temporal precision exhibited by these models. Likewise Akgol et al. used several regression methods, such as SVM, MLP, and M5P, to forecast the uniformity of TCP/IP traffic by comparing the mean absolute percentage error (MAPE) as shown in [71]. The same type of traffic was studied by Cortez et al. in [72], where the dataset was obtained from internet service providers and included various prediction horizons contrasted with statistical approaches.

2.4 Summary

Chapter 2 offered an extensive exploration of network traffic monitoring and analysis (NTMA), covering its fundamental principles, existing applications, and state-of-the-art techniques. The difficulties and factors to be considered for NTMA are elaborated upon, elucidating the requirement of prediction and classification algorithms for network traffic. Some of the key takeaways are given next:

- NTMA is essentially applied in real-time monitoring and management of networking devices to gain insights and control over network performance.

- Some existing NTMA applications include Paessler PRTG Network Monitor and SolarWinds NetFlow Traffic Analyzer.

- Challenges in NTMA include the absence of predictive and classification algorithms for network traffic.

- NTMA is effective in managing bandwidth requirements and detecting cyber-threats.

- Machine learning algorithms are becoming more prevalent in NTMA and are thus being integrated to provide improved predictive capabilities and anomaly detection in network data.

In the next chapter, an in-depth break down of Video Quality Assessment (VQA) is given through its fundamental principles and practical implementations.

2.5 References – Chapter 2

1.   P. Joshi, A. Bhandari, K. Jamunkar, K. Warghade, and P. Lokhande, "IJARCCE Network Traffic Analysis Measurement and Classification Using Hadoop," *International Journal of Advanced Research in Computer and Communication Engineering*, vol. 5, no. 3, 2016, doi: https://doi.org/10.17148/IJARCCE.2016.5360.

2.   Tulsi Pawan Fowdur, L. Babooram, M. Indoonundon, and M. N.-U.-D. I. N. Rosun, *Real-Time Cloud Computing and Machine Learning Applications*, New York: Nova Science, 2021.

3.   A. Siswanto, A. Syukur, E. A. Kadir, and Suratin, "Network Traffic Monitoring and Analysis Using Packet Sniffer," IEEE Xplore, pp. 1–4, Apr. 2019, doi: `https://doi.org/10.1109/COMMNET.2019.8742369`.

4.   J. Bulao, "How Much Data Is Created Every Day in 2021?" TechJury, Jun. 24, 2020, `https://techjury.net/blog/how-much-data-is-created-every-day` (accessed Oct. 29, 2023).

5.   A. D'Alconzo, I. Drago, A. Morichetta, M. Mellia, and P. Casas, "A Survey on Big Data for Network Traffic Monitoring and Analysis," *IEEE Transactions on Network and Service Management*, vol. 16, no. 3, pp. 800–13, Sep. 2019, doi: `https://doi.org/10.1109/tnsm.2019.2933358`.

6.   M. Trevisan, A. Finamore, M. Mellia, M. Munafo, and D. Rossi, "Traffic Analysis with Off-the-Shelf Hardware: Challenges and Lessons Learned," *IEEE Communications Magazine*, vol. 55, no. 3, pp. 163–69, Mar. 2017, doi: `https://doi.org/10.1109/mcom.2017.1600756cm`.

7.   P. Casas, A. D'Alconzo, T. Zseby, and M. Mellia, "Big-DAMA," Proceedings of the 2016 workshop on Fostering Latin-American Research in Data Communication Networks, Aug. 2016, doi: `https://doi.org/10.1145/2940116.2940117`.

8.   M. Abbasi, A. Shahraki, and A. Taherkordi, "Deep Learning for Network Traffic Monitoring and Analysis (NTMA): A Survey," *Computer Communications*, vol. 170, pp. 19–41, 2021, doi: `https://doi.org/10.1016/j.comcom.2021.01.021`.

9.   P. Goyal and A. Goyal, "Comparative Study of Two Most Popular Packet Sniffing Tools-Tcpdump and Wireshark," IEEE Xplore, Sep. 01, 2017, `https://ieeexplore.ieee.org/abstract/document/8319360`.

10.   M. Čermák, Tomáš Jirsík, and M. Laštovička, "Real-time Analysis of NetFlow Data for Generating Network Traffic Statistics using Apache Spark," Veřejné služby Informačního systému (Masarykiana Brunensis Universitas), Apr. 2016, doi: `https://doi.org/10.1109/noms.2016.7502952`.

11.   Risto Vaarandi, Bernhards Blumbergs, and M. Kont, "An Unsupervised Framework for Detecting Anomalous Messages from Syslog Log Files," Network Operations and Management Symposium, Apr. 2018, doi: `https://doi.org/10.1109/noms.2018.8406283`.

12.   R. Munadi, Fardian, Z. Falmiza, M. Ernita Dewi, and Roslidar, "The Performance Analysis of Wireless Distribution System Using Point to Multipoint Network Topology," *MATEC Web of Conferences*, vol. 218, p. 03022, 2018, doi: `https://doi.org/10.1051/matecconf/201821803022`.

13. "Understanding Latency, Packet Loss, and Jitter in Network Performance," Kentipedia, Aug. 08, 2023, `https://www.kentik.com/kentipedia/understanding-latency-packet-loss-and-jitter-in-networking/` (accessed Oct. 29, 2023).

14. A. Abrahams, "Best Practices for Data Cleaning and Preprocessing," `www.jumpingrivers.com`, Aug. 17, 2023, `https://www.jumpingrivers.com/blog/best-practices-data-cleaning-r/` (accessed Oct. 29, 2023).

15. N. Buhl, "Mastering Data Cleaning & Data Preprocessing for Machine Learning," encord.com, Aug. 09, 2023, `https://encord.com/blog/data-cleaning-data-preprocessing/` (accessed Oct. 29, 2023).

16. Constantin Lucian Aldea, Razvan Bocu, and Robert Nicolae Solca, "Real-Time Monitoring and Management of Hardware and Software Resources in Heterogeneous Computer Networks through an Integrated System Architecture," *Symmetry*, vol. 15, no. 6, p. 1134, May 2023, doi: `https://doi.org/10.3390/sym15061134`.

17. W. Jiang, "Graph-based Deep Learning for Communication Networks: A Survey," *Computer Communications*, vol. 185, pp. 40–54, Mar. 2022, doi: `https://doi.org/10.1016/j.comcom.2021.12.015`.

18. A. Nunez, J. Ayoka, M. Z. Islam, and P. Ruiz, "A Brief Overview of Software-Defined Networking," arXiv (Cornell University), Jan. 2023, doi: `https://doi.org/10.48550/arxiv.2302.00165`.

19. A. Kavitha and S. Mary Praveena, "Deep Learning Model for Traffic Flow Prediction in Wireless Network," *Automatika*, vol. 64, no. 4, pp. 848–57, Jun. 2023, doi: `https://doi.org/10.1080/00051144.2023.2220203`.

20. P. Casas, F. Soro, J. Vanerio, G. Settanni, and Alessandro D'Alconzo, "Network Security and Anomaly Detection with Big-DAMA, a Big Data Analytics Framework," 2017 IEEE 6th International Conference on Cloud Networking (CloudNet), Sep. 2017, doi: `https://doi.org/10.1109/cloudnet.2017.8071525`.

21. A. Oliner, A. Ganapathi, and W. Xu, "Advances and Challenges in Log Analysis-ACM Queue," queue.acm.org, Dec. 20, 2011, `https://queue.acm.org/detail.cfm?id=2082137` (accessed Oct. 29, 2023).

22. Amin Shahraki, M. Abbasi, Md. Jalil Piran, M. Chen, and S. Cui, "A Comprehensive Survey on 6G Networks: Applications, Core Services, Enabling Technologies, and Future Challenges," ArXiv, vol. abs/2101.12475, 2021, accessed Oct. 29, 2023. [Online], `https://api.semanticscholar.org/CorpusID:231728460`

23.   A. Shahraki, M. Abbasi, A. Taherkordi, and A. D. Jurcut, "Active Learning for Network Traffic Classification: A Technical Study," *IEEE Transactions on Cognitive Communications and Networking*, vol. 8, no. 1, pp. 422–39, Mar. 2022, doi: `https://doi.org/10.1109/TCCN.2021.3119062`.

24.   J. Liu, J. Wang, Y. Tian, F. Qi, and G. Chen, "Unknown Traffic Recognition Based on Multi-Feature Fusion and Incremental Learning," *Applied Sciences*, vol. 13, no. 13, p. 7649, Jun. 2023, doi: `https://doi.org/10.3390/app13137649`.

25.   S. Zehra et al., "Machine Learning-Based Anomaly Detection in NFV: A Comprehensive Survey," *Sensors*, vol. 23, no. 11, p. 5340, Jan. 2023, doi: `https://doi.org/10.3390/s23115340`.

26.   M. Goldstein, "Special Issue on Unsupervised Anomaly Detection," *Applied Sciences*, vol. 13, no. 10, pp. 5916–5916, May 2023, doi: `https://doi.org/10.3390/app13105916`.

27.   I. Lohrasbinasab, A. Shahraki, A. Taherkordi, and A. Delia Jurcut, "From Statistical- to Machine Learning-Based Network Traffic Prediction," *Transactions on Emerging Telecommunications Technologies*, vol. 33, no. 4, Nov. 2021, doi: `https://doi.org/10.1002/ett.4394`.

28.   A. Alzahrani, T. H. H. Aldhyani, S. N. Alsubari, and A. D. Alghamdi, "Network Traffic Forecasting in Network Cybersecurity: Granular Computing Model," *Security and Communication Networks*, vol. 2022, pp. 1–14, Jun. 2022, doi: `https://doi.org/10.1155/2022/3553622`.

29.   V. Berggren et al., "Artificial Intelligence in Next-Generation Connected Systems," `https://www.ericsson.com/en/reports-and-papers/white-papers/artificial-intelligence-in-next-generation-connected-systems` (accessed Oct. 29, 2023).

30.   E. Ekudden, "Building Cognitive Networks and Human Trust in AI," `www.ericsson.com`, Jun. 30, 2021, `https://www.ericsson.com/en/blog/2021/5/cognitive-networks` (accessed Oct. 29, 2023).

31.   G. Wikström et al., "6G – Connecting a Cyber-Physical World," `www.ericsson.com`, Feb. 11, 2022, `https://www.ericsson.com/en/reports-and-papers/white-papers/a-research-outlook-towards-6g` (accessed Oct. 29, 2023).

32.   T. Keary, "The Best Network Monitoring Tools of 2023," Comparitech, Jun. 30, 2018, `https://www.comparitech.com/net-admin/network-monitoring-tools/` (accessed Oct. 29, 2023).

33.   Solarwinds, "IT Management Software & Monitoring Tools | SolarWinds," Solarwinds.com, 2016, `https://www.solarwinds.com/` (accessed Oct. 29, 2023).

34. "5 Best Network Traffic Monitoring Tools," SolarWinds, Oct. 26, 2021, `https://logicalread.com/network-traffic-monitoring/` (accessed Oct. 29, 2023).

35. "Paessler AG - The Monitoring Company - Producer of PRTG," Paessler, 2018, `https://www.paessler.com/` (accessed Oct. 29, 2023).

36. "11 Best Network Traffic Analyzers For Windows, Mac & Linux," Software Testing Help, Oct. 12, 2023, `https://www.softwaretestinghelp.com/top-network-traffic-analyzers/` (accessed Oct. 29, 2023).

37. Wireshark Foundation, "Wireshark," Wireshark.org, 2016, `https://www.wireshark.org/` (accessed Oct. 29, 2023).

38. I. Shakeel, "Network Traffic Analysis Using Wireshark," AT&T Cybersecurity, Sep. 23, 2021, `https://cybersecurity.att.com/blogs/security-essentials/network-traffic-analysis-using-wireshark` (accessed Oct. 29, 2023).

39. ManageEngine, "NetFlow Traffic Analyzer | Real-Time NetFlow Analysis – ManageEngine NetFlow Analyzer," Manageengine.com, 2019, `https://www.manageengine.com/products/netflow/` (accessed Oct. 29, 2023).

40. "Network Traffic Analysis | Network Analysis and Troubleshooting: OpManager NetFlow Add-on," www.manageengine.com, `https://www.manageengine.com/network-monitoring/network-traffic-analysis.html` (accessed Oct. 29, 2023).

41. "Network Monitoring Software | Network Performance Monitoring Tool: Site24x7," www.site24x7.com, `https://www.site24x7.com/network-monitoring.html`

42. Prometheus, "Prometheus - Monitoring System & Time Series Database," prometheus.io, `https://prometheus.io/` (accessed Oct. 29, 2023).

43. G. Inc, "Network Performance Monitoring Reviews 2022 | Gartner Peer Insights," Gartner, `https://www.gartner.com/reviews/market/network-performance-monitoring`.

44. ManageEngine, "Network Monitoring Software by ManageEngine OpManager," ManageEngine OpManager, `https://www.manageengine.com/network-monitoring/challenges-of-network-monitoring.html` (accessed Oct. 29, 2023).

45. "Top 8 Challenges in Network Monitoring - Forum - THWACK EMEA - THWACK," SolarWinds THWACK Community, `https://thwack.solarwinds.com/resources/thwack-emea/f/forum/6695/top-8-challenges-in-network-monitoring` (accessed Oct. 29, 2023).

46. D. Hein, "7 Network Monitoring Challenges (And How to Overcome Them)," Best Network Monitoring Vendors, Software, Tools and Performance Solutions, Oct. 11, 2019, `https://solutionsreview.com/network-monitoring/7-network-monitoring-challenges-and-how-to-overcome-them/`.

47.   S. Kemp, "Digital 2022: Global Overview Report," DataReportal, Jan. 26, 2022, `https://datareportal.com/reports/digital-2022-global-overview-report` (accessed Oct. 29, 2023).

48.   T. P. Fowdur, B. N. Baulum, and Y. Beeharry, "Performance Analysis of Network Traffic Capture Tools and Machine Learning Algorithms for the Classification of Applications, States and Anomalies," *International Journal of Information Technology*, vol. 12, no. 3, pp. 805–24, Apr. 2020, doi: `https://doi.org/10.1007/s41870-020-00458-0`.

49.   "VNI Complete Forecast Highlights Global Internet Users: Percent of Population Devices and Connections per Capita Average Speeds Average Traffic per Capita per Month Global - 2021 Forecast Highlights IP Traffic," 2016, accessed Oct. 29, 2023. [Online]. Available: `https://www.cisco.com/c/dam/m/en_us/solutions/service-provider/vni-forecast-highlights/pdf/Global_2021_Forecast_Highlights.pdf`.

50.   "Top Web App Development Stats for 2022," SAG IPL - A Technology Blog, Jul. 15, 2020, `https://blog.sagipl.com/top-web-app-development-stats/` (accessed Oct. 29, 2023).

51.   N. M. Balamurugan, M. Adimoolam, M. H. Alsharif, and P. Uthansakul, "A Novel Method for Improved Network Traffic Prediction Using Enhanced Deep Reinforcement Learning Algorithm," *Sensors*, vol. 22, no. 13, p. 5006, Jul. 2022, doi: `https://doi.org/10.3390/s22135006`.

52.   S. Saleem, "Exploring the Future of Cloud Computing in 2020 and Beyond," learn.g2.com, Jan. 14, 2020, `https://learn.g2.com/future-of-cloud-computing` (accessed Oct. 29, 2023).

53.   A. Pradhan, "Network Traffic Classification Using Support Vector Machine and Artificial Neural Network," *International Journal of Computer Applications*, Oct. 2011.

54.   M. Shafiq, X. Yu, A. A. Laghari, L. Yao, N. K. Karn, and F. Abdessamia, "Network Traffic Classification Techniques and Comparative Analysis Using Machine Learning Algorithms," IEEE Xplore, Oct. 01, 2016, `https://ieeexplore.ieee.org/document/7925139`.

55.   M. Dixit, R. Sharma, S. Shaikh, and K. Muley, "Internet Traffic Detection Using Naïve Bayes and K-Nearest Neighbors (KNN) Algorithm," IEEE Xplore, 2019, `https://ieeexplore.ieee.org/abstract/document/9065655/` (accessed Oct. 29, 2023).

56. A. Guezzaz, A. Asimi, A. Mourade, Z. Tbatou, and Y. Asimi, "A Multilayer Perceptron Classifier for Monitoring Network Traffic," *Big Data and Networks Technologies*, pp. 262–70, Jul. 2019, doi: https://doi.org/10.1007/978-3-030-23672-4_19.

57. N. Bayat, W. Jackson, and D. Liu, "Deep Learning for Network Traffic Classification," arXiv.org, Jun. 02, 2021, https://arxiv.org/abs/2106.12693 (accessed Oct. 29, 2023).

58. Z. Wu, Y. Dong, X. Qiu, and J. Jin, "Online Multimedia Traffic Classification from the QoS Perspective Using Deep Learning," *Computer Networks*, vol. 204, p. 108716, Feb. 2022, doi: https://doi.org/10.1016/j.comnet.2021.108716.

59. J. H. Shu, J. Jiang, and J. X. Sun, "Network Traffic Classification Based on Deep Learning," *Journal of Physics: Conference Series*, vol. 1087, p. 062021, Sep. 2018, doi: https://doi.org/10.1088/1742-6596/1087/6/062021.

60. V. Labayen, E. Magaña, D. Morató, and M. Izal, "Online Classification of User Activities Using Machine Learning on Network Traffic," *Computer Networks*, vol. 181, p. 107557, Nov. 2020, doi: https://doi.org/10.1016/j.comnet.2020.107557.

61. J. Bakker, B. Ng, W. K. G. Seah, and A. Pekar, "Traffic Classification with Machine Learning in a Live Network," in 2019 IFIP/IEEE Symposium on Integrated Network and Service Management (IM), 2019, pp. 488–93.

62. A. I. Adekitan, J. Abolade, and O. Shobayo, "Data Mining Approach for Predicting the Daily Internet Data Traffic of a Smart University," *Journal of Big Data*, vol. 6, no. 1, Feb. 2019, doi: https://doi.org/10.1186/s40537-019-0176-5.

63. M. F. Iqbal, M. Zahid, D. Habib, and L. K. John, "Efficient Prediction of Network Traffic for Real-Time Applications," *Journal of Computer Networks and Communications*, vol. 2019, pp. 1–11, Feb. 2019, doi: https://doi.org/10.1155/2019/4067135.

64. J. Zhao and X. He, "NTAM-LSTM Models of Network Traffic Prediction," MATEC Web of Conferences, vol. 355, p. 02007, 2022, doi: https://doi.org/10.1051/matecconf/202235502007.

65. C. Gijón, M. Toril, S. Luna-Ramírez, M. L. Marí-Altozano, and J. M. Ruiz-Avilés, "Long-Term Data Traffic Forecasting for Network Dimensioning in LTE with Short Time Series," *Electronics*, vol. 10, no. 10, p. 1151, May 2021, doi: https://doi.org/10.3390/electronics10101151.

66. J. Sun, "Research on Intelligent Predictive Analysis System Based on Embedded Wireless Communication Network," *Wireless Communications and Mobile Computing*, vol. 2022, pp. 1–11, Feb. 2022, doi: https://doi.org/10.1155/2022/3612073.

67. M. R. Joshi and T. H. Hadi, "A Review of Network Traffic Analysis and Prediction Techniques," arXiv.org, Jul. 27, 2015, `http://arxiv.org/abs/1507.05722` (accessed Oct. 29, 2023).

68. B. L. Dalmazo, J. P. Vilela, and M. Curado, "Performance Analysis of Network Traffic Predictors in the Cloud," *Journal of Network and Systems Management*, vol. 25, no. 2, pp. 290–320, Sep. 2016, doi: `https://doi.org/10.1007/s10922-016-9392-x`.

69. Y. Li, H. Liu, W. Yang, D. Hu, X. Wang, and W. Xu, "Predicting Inter-Data-Center Network Traffic Using Elephant Flow and Sublink Information," *IEEE Transactions on Network and Service Management*, pp. 1-1, 2016, doi: `https://doi.org/10.1109/tnsm.2016.2588500`.

70. S. Gowrishankar and P. S. Satyanarayana, "A Time Series Modeling and Prediction of Wireless Network Traffic," *International Journal of Interactive Mobile Technologies (iJIM)*, vol. 3, no. 1, p. 53, Nov. 2008, doi: `https://doi.org/10.3991/ijim.v3i1.284`.

71. D. Akgol and M. Akay, "Network Traffic Forecasting Using Machine Learning and Statistical Regression Methods Combined with Different Time Lags," *International Journal of Advances in Electronics and Computer Science*, vol. 3, no. 10, pp. 2393–835, Oct. 2016.

72. P. Cortez, M. Rio, M. Rocha, and P. Sousa, "Internet Traffic Forecasting using Neural Networks," The 2006 IEEE International Joint Conference on Neural Network Proceedings, 2006, doi: `https://doi.org/10.1109/ijcnn.2006.247142`.

CHAPTER 3

Video Quality Assessment

In this chapter, the principles of video quality assessment (VQA) and its application on common streaming platforms are unraveled. The fundamental tenets and highlights of VQA are elaborated with regard to maintaining a seamless multimedia experience. The breakdown includes a deep dive into the different artifacts that the algorithm uses, such as ringing, blocking, and noising. With the technique used being a no-reference (NR) metric, a mean opinion score (MOS) is generated, which is interpreted as the quality of service (QoS) rating. This chapter focuses on the building blocks used to construct the VQA algorithm, as well as a review of current applications being used by telecommunications companies and operators. Additionally, an in-depth review of current VQA-based research is presented.

3.1 VQA Fundamentals

The power to summon myriad types of multimedia content with just a gentle swipe of one's fingertips has led to the continuously increasing demand for streaming services such as YouTube, Netflix, Amazon Prime, Twitch, and Vimeo over the last decade [1]. The emergence of over-the-top (OTT) streaming platforms, driven by the effortless integration of smartphones into everyday life, has created a growing need to assess the QoS offered by these major technology firms, along with the quality of experience (QoE) for end users. There is thus a growing complexity around the management of QoS, QoE, expansion, and operational efficiency, driven by the imminent increase in the number of internet applications [2].

Factors that interfere with the effectiveness of multimedia communication via the internet include delays, transmission errors, packet failures, bandwidth limitations, and jitter. These factors significantly impact the signal reception quality, resulting in an unsatisfactory viewing experience for the consumer [3]. The objective is to evaluate the network needs to ensure consistent performance for multimedia applications, sessions,

© Tulsi Pawan Fowdur, Lavesh Babooram 2024
T. P. Fowdur, L. Babooram, *Machine Learning For Network Traffic and Video Quality Analysis*,
https://doi.org/10.1007/979-8-8688-0354-3_3

and network environments, regardless of the impediments. Given that streaming video traffic constitutes a significant portion of current network traffic, internet providers are shifting their focus to developing VQA algorithms, powered with machine learning (ML), to accurately evaluate and replicate the behavior of end users [4].

Gluing together network traffic monitoring and analysis (NTMA) and VQA for network administration brings about a revolutionary change in the functioning of Service Operations Center (SOC) and Network Operations Center (NOC) teams. This enables them to have a comprehensive overview of their networks and to respond promptly [5]. By integrating ML techniques such as deep neural networks, support vector machines, multilayer perceptron, and convolutional neural networks, the QoS analysis for real-time multimedia traffic can be significantly improved beyond what is currently achievable, despite the existence of various software and methods for network traffic management and VQA. Considering the widespread use of real-time multimedia applications on many platforms, including computers and mobile phones, it is crucial to have widespread availability of VQA applications. This is to ensure flexibility and meet the requirements for real-time analysis. This highlights that analytical and prediction methods must evaluate both video quality and the network's QoS in a usual online video streaming situation. Figure 3-1 illustrates a popular architecture around which VQA studies are generally based [6].

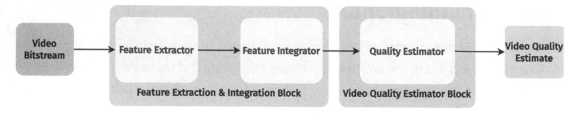

Figure 3-1. *VQA research method [6]*

Video quality is one of the strongest determinants in defining consumer experience and fulfillment in the vast arena of multimedia consumption [7]. VQA is a pioneering methodology that ensures the visual content we interact with satisfies the criteria and expectations of the audience. It serves as the foundation for the pursuit of optimal visual fidelity, with the primary objective of addressing a basic question. What constitutes a visually gratifying and captivating video? It explores the complex relationship between the objective attributes that characterize online videos and the subjective subtleties of human perception. VQA is not an independent entity; it relies on benchmarks, statistics, and defined techniques for its success. The examination of these aspects aims to comprehend the meticulous process of testing and validating the quality of video content [8]. This section explores the fundamental features of VQA, examining the complex framework that encompasses our understanding and evaluation of video quality. Gaining knowledge about the elements that contribute to video quality, the intricacies of human perception, and the techniques used in both objective and subjective evaluations forms the basis of a thorough investigation.

3.1.1 Video Quality Metrics

VQA revolves around a wide array of measurements and evaluated methods exclusively developed to measure the visual accuracy of digital video. These metrics aim to provide a regulated standard for quantifying components of video quality, facilitating unbiased comparisons and evaluations. As a subset of the computer vision realm, VQA seeks to replicate the subjective view of human experience when watching videos, with the aim of obtaining a score that signifies perceptual quality [9]. Some popular benchmarks in the modern era include Live Video Quality Challenge (LIVE-VQC), YouTube User Generated Content (YouTube-UGC), Konstanz natural video database (KoNViD-1k), and Large-Scale Social Video Quality (LSVQ) [10]. On the same wavelength, strict compliance with global benchmarks, namely those set by the International Telecommunication Union (ITU), enhances the assessment procedure [11, 12]. Table 3-1 gives an overview of popular VQA metrics, score systems, and standards used over the years [13, 14, 15].

Table 3-1. *Video Quality Metrics, Score Systems, and Standards*

VQA Metric	Score System	Description
Objective Video Quality Metrics	Peak Signal-to-Noise Ratio (PSNR)	Quantifies the ratio between the highest achievable power of a signal and the power of interfering noise.
	Structural Similarity Index (SSIM)	Assesses the degree of similarity in structure between the original and deformed images, accounting for factors such as brightness, contrast, and overall structure.
	Multi-scale SSIM (MS-SSIM)	Supersedes the SSIM paradigm by outperforming it. This occurs through a frame-by-frame application on the luminance component, after which an average index is calculated.
	Mean Opinion Score (MOS)	A subjective metric converted into an objective numerical value, typically based on evaluations by humans regarding video quality.
Subjective Video Quality Metrics	Single Stimulus Continuous Quality Evaluation (SSCQE)	Involves viewers constantly rating the video quality throughout its playback.
	Double Stimulus Continuous Quality Evaluation (DSCQE)	Enables viewers to continuously rate the quality of a reference video in comparison to the video being evaluated.
Hybrid Metrics and Score Systems	Video Quality Experts Group (VQEG)	Uses a blend of subjective and objective parameters to thoroughly evaluate video quality.
	Hybrid Video Quality Metric (HVQM)	Integrates both subjective and objective indicators to offer a comprehensive assessment of video quality.

(continued)

Table 3-1. (*continued*)

VQA Metric	Score System	Description
ITU Standards for Video Quality	ITU-R BT.500 Series	Describes techniques for evaluating the subjective quality of television images.
	ITU-T Rec. P.910	Explores the use of subjective analysis to evaluate video quality in multimedia applications [16].
	ITU-T Rec. P.913	Details the approaches used to evaluate the subjective perception of video quality, audio quality, and audiovisual performance of internet video and television broadcast quality in any condition [17].
	ITU-T Rec. P.1204	Addresses the subjective VQA of streaming services over reliable transport for resolutions up to 4K [18].

When exploring these measurements, it is clear that a comprehensive evaluation of video quality can be achieved by combining objective, subjective, and hybrid methodologies, both at the pixel level and by catering for human perception. Furthermore, compliance with ITU standards guarantees a universally acknowledged and uniform method, with these standards established for researchers, engineers, and industry experts. According to the ITU, despite the advantages and constraints of every single method mentioned, it is impractical to clearly endorse any individual method over the rest. It is up to the researcher to choose the procedures that are best suited to the given conditions [12].

3.1.2 Human Perception in Video Quality

The assessment of video quality goes beyond the mere technical characteristics of pixels and frames, instead requiring the exploration of the complex regions of human perception. This section examines the intricate interaction between human cognition and visual content to gain insight into how humans perceive and interpret visual stimuli. Some of the psychological and physiological variables are as follows [19, 20, 21]:

- **Cognitive Load**: Refers to the mental exertion needed for information processing. The sense of the video quality can be affected by a high cognitive load, as viewers may experience difficulty in processing intricate visual situations.

- **Attention and Focus**: Interruptions or visual chaos may directly impact the perceived quality, along with the viewer's capacity to concentrate on pertinent components of the frame.

- **Color Vision**: The human visual system is capable of perceiving a wide range of colors, resulting in the perceived quality's being affected if any deviation in color accuracy is observed.

- **Temporal Sensitivity**: The ability to perceive changes over time. Higher frame rates contribute to smoother motion, aligning more closely with the natural temporal sensitivity of the human visual system.

- **Spatial Consistency**: Inconsistencies or distortions in spatial relationships can disrupt the perceived quality of video content.

- **Temporal Consistency**: Refers to the ability to detect and interpret alterations that occur across a time period. Increased frame rates enhance the fluidity of motion, precisely matching the inherent temporal sensitivity of the human visual system.

- **Prolonged Viewing**: Prolonged exposure to multimedia materials may result in eye strain, which can impair the viewer's ability to comprehend high-quality visuals.

- **Engagement and Immersion**: An enthralling and immersive visual experience heightens the perceived level of excellence, cultivating a feeling of involvement and pleasure.

3.1.3 Video Quality Attributes

As the world thrives on multimedia, ensuring that service providers meet the quality standards expected by their subscribers is a direct determinant of success. Countries like China and India exhibit low internet access rates, yet demonstrate high levels of mobile usage for online video streaming. In contrast, Western nations such as the

United States, together with certain regions of Western Europe, have set the bar high for television networks. On the same note, viewers are increasingly opting for video streaming providers due to the availability of more options and the simplicity and leisure of changing services at the click of a button. This is backed up by the 55 percent increase in growth rate of the global streaming market [22]. End users all over the world yearn for top-notch experiences. Based on a poll performed by PricewaterhouseCoopers (PwC) in October 2019, which included over 1,000 individuals aged eighteen to sixty-four in the United States, there is a growing significance placed on simplifying the process of streaming content [23]. The correlation between user-friendliness and dependability is directly related to user involvement and is crucial for promoting long-term usage and favoritism toward video services. Figure 3-2 illustrates the significance of video streaming quality attributes, which are experienced by Akamai's Content Delivery Network (CDN) for video distribution [24].

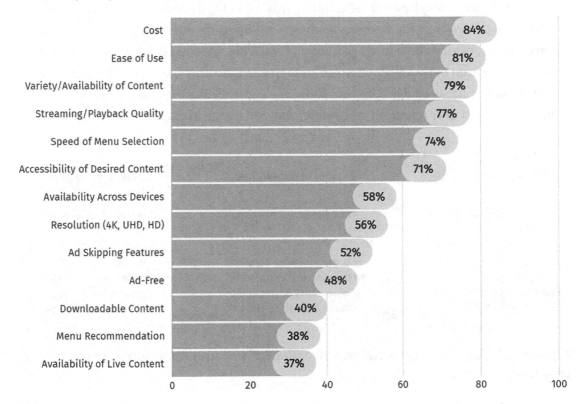

Figure 3-2. *Video streaming attributes rated on importance for Akamai's CDN [24]*

3.1.4 The Optimal VQA Strategy

As viewers progressively consume episodic long-form IP-based material, they have higher expectations for premium video content. The level of excellence of the content is expected to be consistent, regardless of the device or method used. This becomes complicated when the widespread use of mobile devices for watching videos fails to cohere with the fact that audiences in developed countries prefer to watch lengthy serial material and movies on their television screens. According to Vox, Netflix asserts that a significant majority, specifically 70 percent, of their streaming content is consumed on Smart TVs as opposed to mobile phones, tablets, or PCs [25]. Viewers anticipate that both broadcast television and over-the-top (OTT) video will provide a consistent level of quality across all devices [26].

The bottom line is that there is no industry standard that addresses how an ideal video should look. Despite established standards for VQA, there are no codes of practice for delivering video over the internet, leading to no mutual understanding. Thus, video service providers focus on a range of fundamental KPIs, such as rebuffering ratio, video start time, the time between when a user requests a video and it actually plays, and the bitrate, typically measured in Kbps or Mbps [27]. Figure 3-3 shows the link between video quality and user retention, as usually analyzed by the average video service provider.

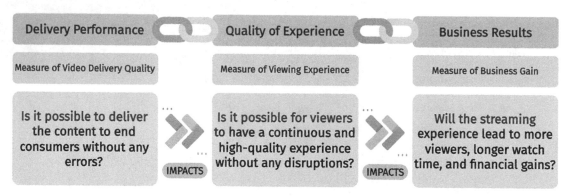

Figure 3-3. *Relationship between video quality and user retention in video service provision*

The path toward standardization consists of first identifying the metrics, the methodology, and the measurement tool so as to maintain accuracy and consistency across the assessment. Figure 3-4 illustrates the addition of two types of metrics to yield the viewer experience [24].

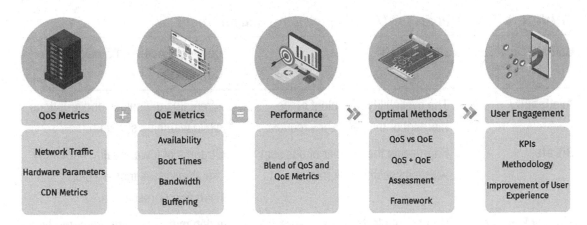

Figure 3-4. *User engagement journey*

With streaming and video quality among the top metrics dictating the success of a video service provider, as shown in Figure 3-2, it is imperative that the design of VQA algorithms be tailored and continuously updated to match the pace of the end user. A robust VQA strategy goes as far as taking into account the customer journey to blend in QoS and QoE metrics.

3.1.5 Quality of Experience (QoE) Metrics

To further understand the necessity of VQA in the digital landscape, it is important to look at the metrics that can be measured at the customer's end. As described by the Customer Technology Association (CTA), *QoE* refers to the level of satisfaction or dissatisfaction experienced by a user when viewing an application or service [28]. The perceived engagement is measured with respect to the level of satisfaction felt by the viewers as a result of the media being consumed, which largely depends on the user's device. QoE metrics can either be objective or subjective. Some prevalent objective QoE indicators typically addressed by video analytics suppliers are given in Table 3-2 [29, 30].

Table 3-2. *QoE Metrics for User Engagement Measurement*

QoE Metric	Description	Consequences of Poor Network Conditions
Availability	The availability of content at the moment of the end user's request.	Streaming difficulties or disruptions while watching streams.
Bitrate	Represents the number of bits sent within a specific time frame.	Streams of lower quality frequently result in videos that are blurry and have poorer resolution.
Playback Failure	Number of playbacks that were terminated due to a playback error following the completion of the initial rebuffering process.	Video may terminate prematurely or the player/browser may crash.
Rebuffering	A video delivery delay that occurs when chunks of the video download more slowly than they play.	Video stutters, freezes, black screens, and potential "wheel of death" occurrences.
Startup Time	Duration between the initiation of playback and the starting point of video playback.	Excessive latency in video initiation or prolonged delay when switching between real-time and linear streams
Video Start Error	Frequency at which playback attempts are halted during video initialization prior to displaying the first video frame.	Absence of video stream.
Lag Ratio	Computed as the waiting time divided by the watch time.	Streams buffer and may result in users leaving in case of a high ratio.

While most platforms try to include most of these measures, the framework used to obtain the final metric typically varies, leading to nonhomogeneous measurements and reporting. For instance, rebuffering is quantified by different tools using different formulas, expressing it as a count, duration, or frequency through computations such as mean or percentages. This depends on the manner and timing in which the occurrences happen.

3.1.6 Quality of Service (QoS) Metrics

QoS metrics revolve around components in the delivery chain that make up the service and performance quality. This ranges from content generation and processing to distribution and last-mile delivery. To address the QoE discrepancies and precisely identify the cause of the service degradation, QoS metrics need to be established for every link in the video distribution process, such as the viewer's ISP, the CDN, and the content preparation facilities. Other interconnected elements include availability, throughput, transmission time, and jitter [31]. From a CDN's perspective, video distribution relies on several interconnected elements, as elaborated in Table 3-3 [24].

Table 3-3. *QoS Metrics for Performance Measurement from a CDN's Perspective*

QoS Metric	Description	Factors
Traffic/Hits	Amount of data being transmitted across the CDN, measured in either number of visits or bytes	Source, midgrass, edge hits
Throughput	Capacity available for data transfer across the CDN	Represented in Mbps
Offload	Total number of requests that are not returned to the source	Hits and volume
Availability	Calculated as the ratio of successful hits over the total number of hits	Edge and source responses

3.1.7 Quality of Performance (QoP) Metrics

To achieve a consistent understanding of the viewer's experience, it is necessary to use specific metrics that assess the QoS for each component of the video distribution process. These metrics should be closely aligned with industry standards for assessing the quality of experience. QoP indicators are characteristics that outline operational data to enhance system performance and offer insight into business elements by acting as proxy QoE measures. If a customer is viewing video segments with a duration of six seconds, and the CDN exceeds six seconds in delivery of the segment, the client player would experience rebuffering. This can be measured using a QoP metric, which also functions as a proxy parameter for customer-end rebuffering [24].

3.1.8 Subjective VQA

Humans, who are the ultimate consumers, serve as the definitive evaluators of visual quality, where judgments can be acquired through subjective experiments. This entails the participation of a group of individuals, typically non-experts, who are commonly known as test subjects, with the role of evaluating the perceptual quality of specific test material, such as a series of movies [32]. Subjective experiments are commonly carried out in a controlled laboratory setting, as detailed in the ITU-R BT.500 recommendation [12]:

- Viewing distance.

- Room illumination.

- Normalized protocols.

- Tests duration.

- Observer's recruitment.

- Methods to detect incoherent observers and reject their scores.

- Methods to compute the precision of the MOS.

Prior to conducting a subjective experiment in a controlled environment, it is necessary to engage in meticulous planning and take into account the evaluation method, selection of test content, viewing circumstances, grading scale, and time of presentation. The different types of methods for subjective VQA are as follows [33]:

- **Single Stimulus**: The participant is presented with different versions of the test content without any reference for comparison. One such example is the Absolute Category Rating (ACR) method.

- **Double Stimulus**: By being exposed to two videos consisting of the reference one and the degraded one, the participant rates the change in quality between the two feeds. This is known as Degradation Category Rating (DCR).

- **Multi-Stimulus**: The participant judges the quality of several video content samples against the reference video, with the ability to review the footage multiple times. An example is the Subjective Assessment of Multimedia Video Quality (SAMVIQ).

These methods yield individual scores given by the participants, such as MOS or differential mean opinion score (DMOS), depending on the setting of the experiment.

3.1.9 Objective VQA

Given that subjective experiments require intensive planning, physical space, and time, researchers and engineers have turned toward objective quality measurements, which aim to produce the same MOS accurately. Two different procedures are explained as follows [34]:

- **Psychological Metrics**: These parameters seek to represent the human visual system by incorporating factors such as contrast and motion awareness, frequency selection, spatial and temporal patterns, masking, and color judgment. These metrics are applicable to a broad range of video degradations, although their computation is typically resource-intensive and not commonly employed in streaming applications.

- **Engineering Metrics**: By first extracting specific elements or artifacts from a video, these are parameters that do not overlook the qualities of the human visual system. Nevertheless, the underlying notion behind their design is to analyze video content, rather than focusing on fundamental vision modeling. A collection of characteristics is then molded into an objective approach to assess quality, which is in turn correlated to predict the MOS.

Based on the environment under which the assessment is to be performed, objective approaches can be categorized into the following methods [35, 36]:

- **Full Reference (FR)**: This method allows for the complete original video to be accessible for reference. FR approaches rely on the comparison of an altered video with the original footage.

- **Reduced Reference (RR)**: In this method, representative characteristics of the original video are provided to the algorithm and the insights from the test video are compared against this matching information.

- **No Reference (NR)**: With no access to the original video, this approach uses several methods such as considering artifacts within the pixel domain of the test video, using details rooted in its bitstream, or designing a hybrid pixel-based and bitstream-based model.

3.1.10 Quality Metrics for Network, Video, and Streaming

After analyzing the QoE, QoS, and QoP metrics of an objective nature, this section elaborates on the subjective notion of end users when it comes to video quality. A summarized version of these metrics is given in Figure 3-5 [37].

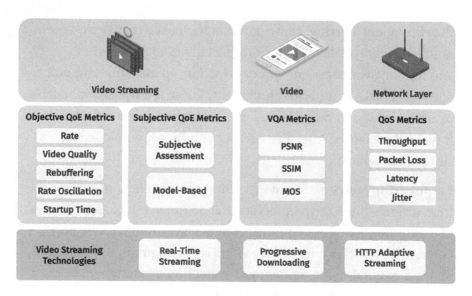

Figure 3-5. *Quality metrics for network, video, and streaming [37]*

3.1.11 Video Quality Databases and Benchmarking

Video quality databases and benchmarking are essential for the progress of VQA as they act as continuous test beds and datasets for computational evaluations. Some video quality databases are as follows:

- **LIVE Video Quality Challenge Database (LIVE-VQC)**: Consists of an extensive repository of videos that includes a wide range of resolutions and content categories [38].

- **TID2013**: Designed to address distorted photos and videos, provides a wide range of difficulties for algorithmic evaluations. The content incorporates deliberate degradations, replicating real-life situations where there is a deterioration in quality [39].

- **YouTube UGC dataset (YouTube-UGC)**: Designed to assist in the progress of research on UGC video quality evaluation and video compression [40].

- **Netflix Video Multimethod Assessment Fusion (VMAF) Dataset**: Created by Netflix, this dataset encompasses a broad spectrum of material, including videos with different levels of complexity. It is highly pertinent for evaluating VQA algorithms tailored for streaming services [41].

On the same wavelength, some benchmarking methodologies are presented next [42]:

- **Standardized Testing Protocols**: Benchmarking approaches are used to build uniform assessment protocols that guarantee consistency and reproducibility. These guidelines establish the criteria for testing algorithms, ensuring unbiased and comparable evaluations.

- **Objective Metrics as Reference Points**: With methods such as the PSNR and SSIM, these metrics offer measurable standards for evaluating the precision of VQA models.

Likewise, the prominent challenges and considerations concerning VQA datasets are listed here:

- **Scalability**: New displays with groundbreaking advancements and video resolutions hit the market every year, requiring VQA models to be constantly up-to-par in performance.

- **Realism and Diversity**: Viewing conditions have to be homogeneous in all testing environments to remove bias.

- **Temporal and Spatial Variability**: Video content contains variations in temporal patterns and spatial complexities, which pose as obstacles for VQA algorithms, requiring the model to be more complex and effectively adapt to varied circumstances.

3.1.12 Temporal and Spatial Considerations in VQA

Temporal factors in VQA pertain to the changes in content over time and their influence on viewer perception. Spatial issues, meanwhile, apply to the stationary characteristics of specific frames. Table 3-4 discusses the types of attributes to consider in the VQA realm [43, 44].

Table 3-4. *Types of Attributes to Consider in the VQA Realm*

Consideration	Attribute	Description
Temporal	Frame rate and motion	An increase in frame rate contributes to motion smoothness, reduced occurrence of motion blur, and a boost in the clarity of visuals.
	Temporal consistency	Fluctuations or irregularities in the frame rate can disturb the viewer's sense of immersion, resulting in an apparent drop in quality.
	Dynamic scene changes	Frequent and abrupt scene transitions require algorithms to be flexible to sudden changes in content.

(continued)

Table 3-4. (*continued*)

Consideration	Attribute	Description
Spatial	Resolution and detail	A higher resolution signifies a greater level of detail, enhancing the level and perceived quality of visualizations.
	Compression artifacts	The identification of artifacts such as blurring and blocking is paramount for VQA models.
	Aspect ratio and composition	Aligning the spatial features with aesthetic norms enhances the viewing experience.
Evolution	Adaptive bitrate streaming (ABS)	Network conditions go hand in hand with the performance of streaming services, where ABS already bridges the gap between available bandwidth and streaming quality.
	ML for Temporal Analysis	Finding patterns in movement, changes in scenes, and fluctuations in frame rates.
	Spatial Compression Techniques	Approaches such as High Efficiency Video Coding (HEVC) can be used to enhance spatial compression efficiency.

3.1.13 VQA for Evolving Video Content

The video content industry is ever-evolving, with new formats, platforms, features, and the increasing popularity of user-generated content (UGC) [45]. Conventional VQA methodologies may not always be easily adapted to these new content formats, such as in the case of UGC where users use smartphones from an extremely large pool of devices, which adds to the intricacies of assessing video quality [46]. This section investigates the approaches and requirements of new VQA models, as follows:

- **Short-form Videos**: As one of the most engaging types of multimedia content today, platforms such as TikTok have introduced short videos that captivate the attention of users, with their own algorithms that recommend videos based on the watch history. The immense plethora of content genres, editing styles, and devices used to record the videos needs to be thoroughly addressed by robust VQA algorithms [47].

- **Interactive and Immersive Content**: Extending beyond standard linear formats, virtual reality (VR) and augmented reality (AR) incorporate spatial and interactive components, which require VQA to consider how these factors affect the user experience. Some features include 360 video characteristics, projection, encoding pressure, cache pressure, bandwidth pressure, and stalling patterns and variations in encoding quality [48].

- **Quality Variability in UGC**: Videos are filled with quality disparities, ranging from expertly crafted videos to amateur recordings.

- **Subjective Preferences**: Different content genres appeal differently to end users, with types such as sports, movies, comedy, dancing, learning, fitness, food, and trends.

- **Using AI for IPTV**: Through the automation of analysis, optimization of encoding, and feedback enhancement, AI can use ML and computer vision techniques to identify and categorize video quality issues, such as blurring, blocking, or freezing. It can then offer practical recommendations for improving the video delivery [49]. Likewise, the encoding parameters and compression schemes can be enhanced, thus minimizing bandwidth and storage demands, while maintaining video quality. Furthermore, natural language processing (NLP) and sentiment analysis can be used to gather and dynamically analyze comments from viewers as feedback for the level of satisfaction and preferences.

- **Blockchain for IPTV**: Smart contracts and tokens in blockchain can serve as incentives for the VQA process, allowing participants to be rewarded for their input and feedback [49]. Peer-to-peer (P2P) networks can be used for the distribution and streaming of video content, allowing a drop in reliance on central servers and networks, and alleviating congestion. This would help contribute to the QoS of the network.

3.2 Existing VQA Applications

The realm of multimedia is known for its ever-changing nature and growing complexity, requiring continuous effort to improve viewer experiences, even as connectivity increases around the world. Current VQA applications provide solutions that cover a range of services, such as broadcast, streaming, telecommunications, video conferencing, and IPTV, using tailored and advanced algorithms to both measure and guarantee the best possible video quality [50]. This section delves into prominent off-the-shelf software options and analyzes their features, influence, and contributions to the VQA community.

3.2.1 Sentry by Telestream

With a quality graphical user interface (GUI) displaying characteristics of the channels being monitored, Sentry is a software-based quality monitoring product that can be adopted for cable, broadcast, and over-the-top (OTT) streaming services, i.e., streaming services over the internet [51]. The objective of the framework is to monitor the quality of the content being deployed such that any factors affecting video quality can be detected and therefore resolved. One of the main highlights includes reacting to the content delivery quality before the client reacts so as to avoid discontent and complaints. The software caters for both broadcasting and streaming services, on top of offering a picture quality assessment feature that analyzes parameters such as tiling, noise, and blur to produce informative graphical results. Different versions are available depending on the user's requirements and can be programmed to operate on the cloud, on virtual private networks, as well as in data centers. The company uses the custom-made TekMOS ML algorithm, which performs frame-by-frame analysis based on luminance to determine video quality characteristics before producing a QOS score as well as a TekMOS graph.

115

3.2.2 Real-Time Media Assessment (RTMA) by ThinkTel

By first conducting an assessment of the corporate network, RTMA gathers information about network capacity and peak times to generate a full report. The metrics that are taken into consideration include latency, jitter, and capacity with respect to both audio and video services [52]. The goal is to provide the client with recommendations on how to further optimize their network, which consists of data from a multitude of business communications software such as Microsoft Teams. The latter could potentially strain the network resources during peak times, thus affecting the QoS of employees. RTMA therefore provides capacity planning and cost and time savings, and also guarantees an improved user experience. ML capabilities are not included, as the software mainly stresses data gathering.

3.2.3 Witbe

As one of the leading QoE monitors when it comes to IPTV worldwide, Witbe has mastered the concept of automated testing for set-top boxes (STBs), mobile devices, and web browsers through the deployment of robots, which are equipped with VQA capabilities [53]. Their strong cloud model allows video captures to be taken, and the devices are controlled in real time from a distance [54]. The personalized scripting capability allows scenarios to be written and deployed in an instant. For example, streaming services such as Netflix and YouTube can be monitored through the robots, which are equipped with special optimized cameras to perform VQA, thus yielding a proprietary subjective MOS score, together with metrics such as blurring and blocking. Witbe also caters to the gaming realm, with tests performed on Google's cloud gaming platform, Stadia [55]. KPIs such as availability, login, login time, game availability, game launch time, playability, and Witbe's VQ-MOS score are some of the generated metrics, which can then be viewed on Witbe Datalab, where customizable dashboards are found for analysis. Subscribers also have the option to receive immediate notifications of video and audio issues. These alarms are then promptly presented in real time on the Witbe Remote Eye Controller, allowing NOC teams to quickly respond to any discovered issues. Witbe VQ-ID [56], which is Witbe's latest algorithm, additionally calculates the duration of a video that was free from any issues, as well as the time during which it identified significant video and audio deteriorations, such as macro-blocking, blur, and saturation. These metrics provide marketing teams and managers with high-level KPIs. Figure 3-6 illustrates Witbe's architecture [57].

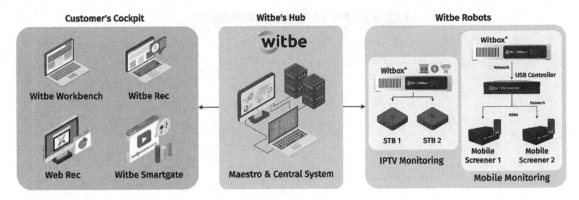

Figure 3-6. *Witbe's architecture [57]*

3.2.4 ViCue Soft

At its core, the VQ Probe available as one of the products of ViCue Soft supports both objective and subjective metrics, together with comparing two videos [58]. There is support for the most well-known metrics, such as PSNR, SSIM, and VMAF, including heat map differences, scene change detection, and zooming down to pixel values. Modern standards such as AVC/H.264, HEVC/H.265, VVC/H.266, AV1, AV2, AVS3, and MJPEG are supported. This tool enables users to investigate and evaluate various codec standards, generate rate-distortion curves, and calculate bit-rate reduction. In the hierarchy view of the software, the information displayed includes the degree of reciprocity of B-frame references, frame numbers in the sequence of decoding and display, the color-coded frame type, and frame references. In the metric view, the analyzer illustrates PSNR and SSIM representations for each component, namely one for luma (Y) and two for chroma (U and V). To output the charts, the full stream can be analyzed, with support for both file-based and real-time analysis. Likewise, the VQ Analyzer targets QA engineers, technical support engineers, and broadcasters, as well as non-professionals, for a number of services, such as broadcasting, eSports, streaming, and social networking. Figure 3-7 illustrates an example of a slightly degraded video being compared against its reference video to generate the VMAF score.

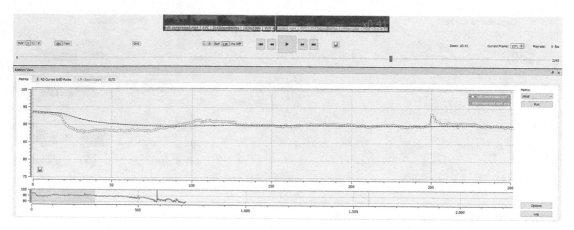

Figure 3-7. *Example of VQ Probe to generate VMAF score for a video*

3.2.5 AccepTV Video Quality Monitor

AccepTV's adaptability is underscored by its compatibility with various video formats, allowing for a seamless integration into different content production and streaming workflows [59]. The primary product, i.e., Video Quality Monitor, excels in measuring video and audio quality, coupled with lip-synchronization from audio and video files. It caters to STBs through HDMI cables; IP streaming protocols such as UDP, RTP, HTTP, HTTPS, and RTMP; as well as desktop capture. Some of the metrics that can be measured are audio quality and volume, lack of depth, choppiness, buffering interruptions, occurrence of black frames, channel switching times, video startup time, presence of noise, excessive brightness, absence of sound, color intensity, and video stability. As an NR method, its MOS ranges from 0 to 100, providing both offline— i.e., file-based—and online usage. It works by analyzing the individual pixels of the decoded frames, together with extracting parameters from the encoded bitstream, yielding a hybrid model. One of its prominent features is the ability to correlate GPS coordinates to match location to field testing of 5G, 4G, and Wi-Fi networks [60].

3.2.6 VQEG Image Quality Evaluation Tool (VIQET)

The Video Quality Experts Group (VQEG) was formed in 1997 to propel the VQA sphere forward and encompasses the ITU-T and ITU-R study groups [61]. Recently geared toward new application areas such as gaming and AR/VR, the group came up with an image quality evaluation tool, thus naming it VIQET [62]. As an open-source tool, it is specifically developed to assess the quality of pictures taken by different devices so as to produce an MOS ranging from 1 to 5 for each one. The software recommends that the same scene be captured by the user, with a similar field of view, on the various test devices. It also contains four broad categories of pictures, such as those in the outdoor setting, indoor wall hangings, indoor arrangements and objects, and outdoor environments. Users can thus choose to input their photos into the preset categories for increased algorithm accuracy. Available at its GitHub repository [63], the VIQET program was tested for an outdoor scene as depicted in Figure 3-8, which also gives the screenshot from a downloadable PDF report available after the test.

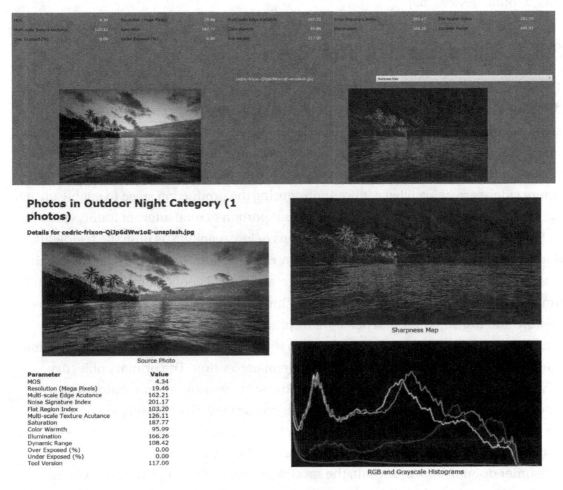

Figure 3-8. *Example of VIQET being used to measure the MOS of an image taken in an outdoor night setting*

3.3 State-of-the-Art Review of VQA

In 2023, the typical individual consumed around seventeen hours of internet video content every week, which dropped from nineteen hours in 2022 [64]. This change in viewing patterns highlights the importance of VQA in satisfying viewers' changing expectations as well as the shifting dynamics of content consumption. To comprehend the advancements and developments that define the present video quality environment, it is crucial to undertake a comprehensive examination of VQA, considering the ongoing changes in the digital realm. This section provides a nuanced comprehension of the cutting-edge approaches that characterize the present age of multimedia communications, where 75 percent of people consume short-form video content on their mobile devices [65].

3.3.1 Background of VQA

The development of VQA stems from the need to synchronize multimedia content distribution with user demands for exceptional visual experiences. The recent advancements in mobile technology have handed the world the incredible superpower of recording videos of cinematographic caliber at any time and location, allowing them to document their everyday experiences. This is the case in Instagram Reels and TikTok, where videos are being filled with captions during the production stage to enable greater accessibility [66]. Despite being a predominant portion of total internet traffic, video traffic undergoes a number of processing steps before reaching its ultimate recipients, during which the perceived video quality may either degrade or be enhanced. VQA sits at the heart of the operation, thus having several practical applications, such as real-time quality monitoring, performance assessment of video architectures such as video capture and transmission, and perceptual optimization of multimedia frameworks. Traditional methods of VQA resorted mainly to the generation of technical information but did not fully capture the subtleties of human perception. The primary objective of VQA research is to create objective methods that have a strong correlation with subjective methods [67]. According to Brightcove, 62 percent of viewers are prone to developing an unfavorable opinion of a brand if they come across video material of poor quality [68]. This figure highlights the pivotal importance of VQA in influencing consumer opinion and, as a result, the success of digital content providers. VQA poses as not only a technical need, but also a strategic necessity for organizations striving to succeed in the competitive realm of multimedia communications.

3.3.2 Machine Learning in VQA

Subjective elements of video quality have slowly seeped through the most robust ML algorithms with their adoption in VQA. Despite facing several roadblocks in terms of perceptual quality evaluation, ML is now extensively used to propel the coding effectiveness of video codecs [69]. ML is further adopted by one of the world's largest tech giants, Netflix. Its ML solution VMAF combines visual information fidelity (VIF) [70] and detail loss metric (DLM) [71] to measure image quality by calculating the temporal difference of the luminance components between consecutive frames on a pixel-wise basis. Support vector regression (SVR) is then used to compute a weighted average, quantifying the ultimate perceptual quality score. Due to its reliance on frame-level features, VMAF does not consider the intricate motion complexity of the video, which surpasses mere pixel-wise variations in subsequent frames. Over the years, notable end-to-end ML and DL approaches have been proposed for VQA, including C3DVQA by Xu et al. [72], and CompressedVQA by Sun et al. [73]. The latter, for example, demonstrated the shear performance of ML models with quality-aware features, using a simple MLP network to predict chunk-level quality scores where multiscale weights were derived from the contrast sensitivity function of the HVS.

Despite the high-flying models powered by ML, enhanced streaming methods amplify the need for more bandwidth capacity and introduce additional intricacies in transmission, both of which are crucial factors for services and internet service providers. Historically, this problem has been examined within the framework of managing network QoS. Nevertheless, streaming via wireless networks experiences a decline in quality, even if there is enough nominal capacity, because of the fluctuating channel conditions [74]. This deterioration cannot be accurately evaluated simply through the incorporation of QoS factors. These variables only provide insight into the condition of specific networks and do not fully include the end-to-end characteristics that impact the overall quality experienced by the user. To tackle these components, the management of QoE has been acknowledged as a much more efficient approach [75]. Newer trends are therefore centered around predictive QoE management through ML, as depicted in Figure 3-9 [76].

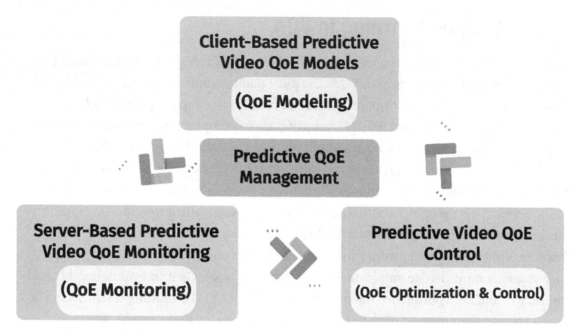

Figure 3-9. *Integration of predictive models in QoE management [76]*

3.3.3 Machine Learning Algorithms to Analyze Video Quality in Multimedia Communications

Prior studies have mostly concentrated on discovering ML algorithms that are well-suited for detecting and responding to features that govern a multimedia communications system, particularly when it comes to internet-based communication. Mohamed et al. replicated and modeled the response of a sample of individuals regarding variations in video quality using ANNs [77]. The factors considered included packet loss rate, loss distribution, bitrate, and frame rate. Given that quality assessment encompasses both objective and subjective standpoints, the study aimed to mitigate bias around subjectivity by emphasizing accuracy. The systems were deployed in scenarios that required video transmission via packet networks. However, the assessment did not take into consideration features such as audio transmission along with video, thus ignoring aspects such as lip synchronization. The ANN model is trained using the subjective data measure from human tests groups, thus acting as an emulator. Moreover, the authors repeated the training procedure to improve the algorithm. Following the subjective evaluations by twenty participants, the ANN model with predetermined distortion parameters was then implemented on the real-time video architecture over packet networks. The method yielded a correlation coefficient of 0.9907 and an average inaccuracy of 0.253, thereby demonstrating the effectiveness of emulating human video quality assessment. A summary of the methodology used is provided in Figure 3-10.

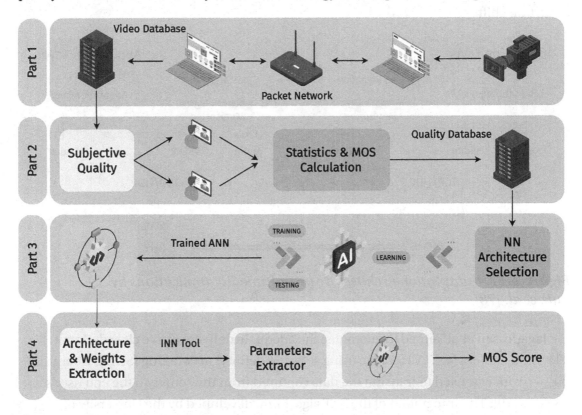

Figure 3-10. *Workflow of the proposed methodology for the MOS score generation by Mohamed et al. [77]*

Ruiz et al. introduced a new approach for effectively optimizing the QoS offered by real-time multimedia applications [78]. Figure 3-11 gives the general adaptation architecture in multimedia applications as presented in the paper. The method included adjusting suitable parameters selected by ML algorithms to optimize the network's performance based on its available resources. The modifications to the parameters are executed on the server side, namely within the adaptive apps. In short, the applications are configured to fine-tune their settings in real time to both suit the needs of the end user and satisfy user-perceived QoS requirements. To account for circumstances including video and audio loss in a mobile ad hoc network prototype, the authors argued that these tests needed to be run in a packetized ecosystem at the network layer. The slipper rule induction algorithm was used to learn which settings would be preferred by the end user, based on scores extracted from real users. A genetic algorithm determines the point at which the adaptive program initiates the settings modification. The authors claimed that their method worked reliably even in situations of persistent distortion shifts.

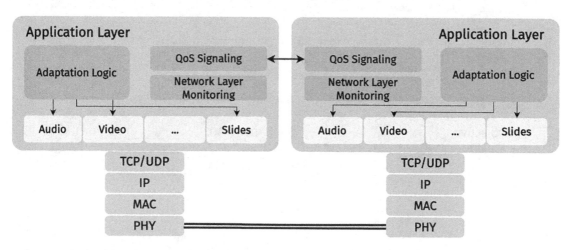

Figure 3-11. *Adaptation architecture in multimedia applications by Ruiz et al. [78]*

Jagadessan et al. aimed to guarantee the adequate delivery of live closed-circuit television camera (CCTV) footage using software defined networking (SDN), where the network was used to transmit the data captured from the source to the end user's displays [79]. The integration of the BAT algorithm, developed by the University of Cambridge in 2010, with an AI module located in the control switch was intended to prioritize data packets. An IoT cloud was used as the database. The experiment resulted in a 70 percent improvement in network performance when it comes to the characteristics of jitter, as compared with a network without SDN.

Thomos et al. conducted a comprehensive analysis of recent advancements aimed at enhancing the quality of multimedia material using machine learning techniques [80]. The paper dives deep into the applications of ML in several realms, such as AR/VR/XR and 360-degree video, video streaming systems, caching, and QoE assessment. For example, in self-driving vehicles, intelligent cars need video material to be delivered at a pace of 750 MB per second, with ultra-low latencies. To achieve this, ML models have been created based on prior research. From another standpoint, multiple characteristics are merged and processed into a final score using SVM regressors. CNNs are also explored for integrating multimodal inputs and enhancing image quality assessment (IQA) by incorporating concepts of the HVS, such as saliency. Instead of maps, saliency features are combined with IQA properties in a hierarchical manner, resulting in the gradual improvement of the latter as the network depth increases [81]. In turn, the combined characteristics are used to estimate the objective picture quality. These kinds of models also allow for effective management of resources, pre-fetching of data, caching services, and improvements in security measures.

Fu et al. presented a sequential reinforcement learning (RL) approach, equipped with a dynamic tiling mechanism to address the high dimensionality of the search space for 360-degree video streaming [82]. The typical tile-based video streaming layout is shown in Figure 3-12. Meanwhile, Kan et al. upscaled the previously cited tiling approach to assign bitrates to each tile through a probabilistic framework [83]. The model was further strengthened by incorporating feedback from the RL agent, enhancing the effectiveness of this method. This research resulted in a significant improvement in 360-degree video streaming concerning the QoE of the end user.

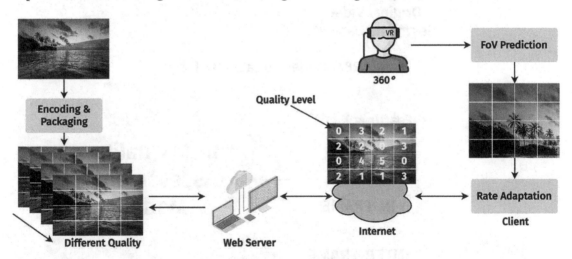

Figure 3-12. *Workflow of the general layout for tile-based streaming by Fu et al. [82]*

Likewise, Vega et al. highlighted the significance of being capable of generating real-time QoE analysis and measurements, even in challenging network conditions [84]. Therefore, they opted for a lightweight NR strategy that leverages client-side variables as input to ML approaches. This strategy aims to attain a high degree of precision, adaptability, and sustainability using Boltzmann machines, in which every node is connected to every other node. The inputs consist of video characteristics that have been retrieved from affected video sources. The methodology underwent benchmarking against both the LIMP and ReTRiEVED video quality statistics using FR techniques. Figure 3-13 shows how restricted Boltzmann machines (RBMs) are coupled with real-time unsupervised and deep learning (UDL)–based VQA. The model reached a minimum precision of 85 percent when considering all available occurrences.

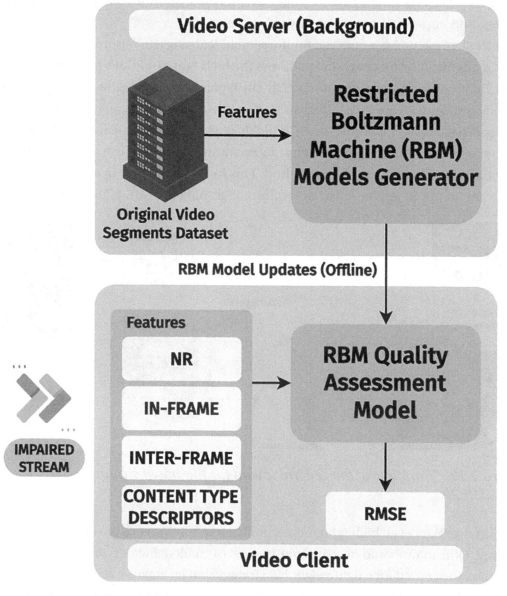

Figure 3-13. *Workflow of the proposed methodology for real-time UDL-based VQA by Vega et al. [84]*

Vega et al. carried out an in-depth review of the techniques used to measure QoE in video streaming services [76]. The frameworks involved predictor methods for gauging QoE in hopes of highlighting the best ML algorithms for this purpose. With respect to client-based forecasting with QoE systems that are centered around the MOS, it has been noted that RR and NR designs outperform FR platforms. This is because FR frameworks require access to the source material, which leads to a time-consuming and intricate algorithm. Nevertheless, the independent RR and NR models proved inadequate in precisely forecasting the human perception of video material. The paper states that ML substantially increases accuracy through the implementation of proactive quality of experience loops for regulation. One of the many predictive frameworks demonstrated in this paper is shown in Figure 3-14. These loops involve monitoring tools at both the consumer and server ends to facilitate immediate adjustments. An example is demonstrated by Konuk et al. [85], where K-means is used to spatially cluster video material based on bitstream and spatio-temporal content descriptors.

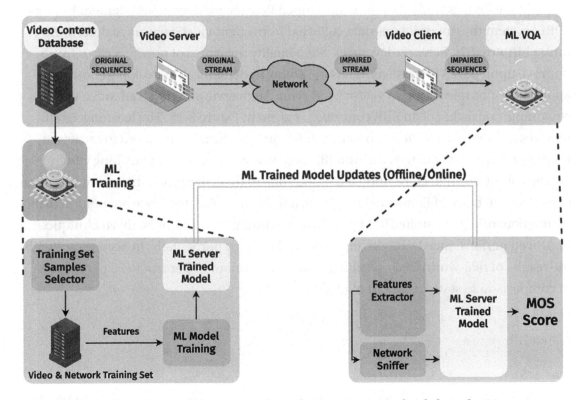

Figure 3-14. *Workflow of the general predictive VQA methodology by Vega et al. [76]*

Meanwhile, Raca et al. focused on investigating how ML and DL approaches may produce precise forecasts for video streaming services on cellular networks [86]. The authors resorted to support vector machine (SVM) and random forests (RF) as the training algorithms for the prediction of throughput. Subsequently, a comparison was produced with the long short-term memory (LSTM) framework in terms of their respective capabilities, training requirements, and data demands. The QoS metrics in a single cell were determined by considering parameters such as the quantity of devices connected and the average throughput. The derived characteristics include bitrate, switch count, average switch count, and the mean bitrate of particular video segments. This predictor was then coupled with a 4G-powered HTTP Adaptive Streaming (HAS) video player, leading to outstanding and anticipatory assessments, in contrast to the reactive nature of a traditional video player. This resulted in significant enhancements in terms of seamless transitions between different levels of quality and rapid adaptation to unexpected falloffs in the available network capacity.

In line with this study, Mao et al. developed Pensieve, a deep reinforcement learning (DRL) system that incorporates data collected from client video viewers to determine future bitrates [87]. Pensieve leverages video quality measures of previous cases to improve upon past ABR verdicts. It has the ability to dynamically develop a control approach for bitrate adaption via trial and error. An overview of HTTP adaptive video streaming, consisting of an ABR controller, is given in Figure 3-15. The learning agent receives as input, in addition to bandwidth factors, the duration required to retrieve a video portion. Consequently, the ABR RL agent was capable of distinguishing which policies outperformed conventional methods due to defective modeling techniques. The overall architecture of Pensieve through the application of RL to bitrate adaptation using neural networks is illustrated in Figure 3-16. According to the authors, this technique improved performance by modifying its control strategy perpetually, in accordance with the results of real-world trials, resulting in an approximate 12 percent to 25 percent rise in comparison to state-of-the-art platforms.

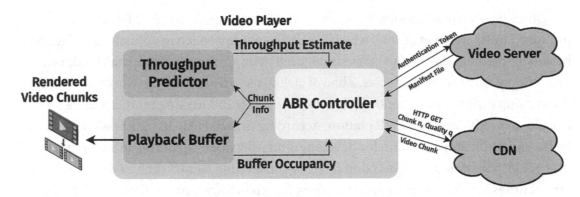

Figure 3-15. *Overview of HTTP adaptive video streaming by Mao et al. [87]*

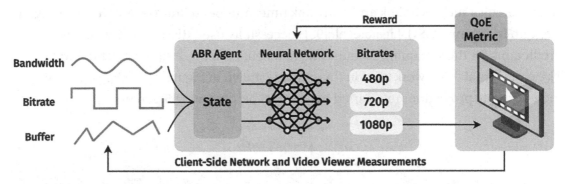

Figure 3-16. *Workflow of the application of RL to bitrate adaptation by Mao et al. [87]*

In comparable research, Yousef et al. used supervised classification to estimate the category of ABR by forecasting its prominent properties based on the average reduction in impurity [88]. The authors claim that this methodology can forecast the bitrate selection of any ABR method by providing a range of input properties that may be used at the application level. The uphill prediction battle was solved by refactoring it into a multiclass classification problem, which consisted of quality of delivery (QoD) metrics, such as buffer size, speed levels, and download times during past and present phases. Several ML methods were compared for their ability to make predictions, including logistic regression, SVMs, RFs, DTs, gradient boosts (GBs), and naïve Bayes (NB). The experiment concluded with the RF and GB models on top for the most effective performance.

Other notable publications in this area of research include the ACTE method, proposed by Bentaleb et al. [89], which is a distinctive approach to adjusting the bitrate based on fragment downloads. This strategy led to improved bandwidth and reduced instances of stutters and latencies. Zhao et al. [90] used two neural networks to enhance the viewing experience by controlling network latency and making more accurate decisions on ABR and delay adaptation. According to Sani et al. [91], RF was shown to be the most efficient algorithm for developing an ABR video streaming infrastructure over HTTP. The evaluation was conducted by considering factors such as download and arrival times for video slices, types of codecs, length of fragments, and bandwidth. Da Hora et al. [92] achieved an RMSE of 0.60 to 0.79 in their attempts to compute the MOS of YouTube videos using support vector regression (SVR). The researchers included characteristics such as uplink and downlink time, number of frames, and received signal strength indicator (RSSI). Ligata et al. [93] successfully used RFs to provide precise predictions for the streaming-to-buffering ratio, buffer delay, frequency, and duration. The conceptual framework used in this paper is shown in Figure 3-17. Consequently, their accurate projections rate varied between 85 percent and 95 percent.

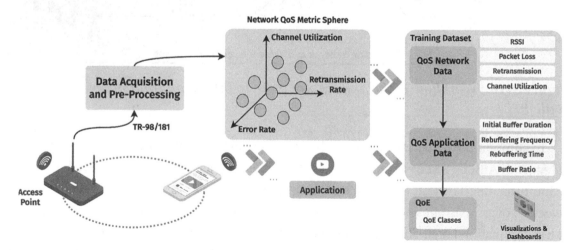

Figure 3-17. *Workflow of the conceptual framework proposed by Ligata et al. [93]*

3.4 Summary

Chapter 3 provided a thorough analysis of video quality assessment (VQA) that encompassed its core principles, existing applications used by industry professionals worldwide, and cutting-edge methodologies. It also walked the user through subjective and objective analysis concepts before emphasizing the need for quality assessment for evolving video content. This is followed by a state-of-the-art review of current VQA trends in the research community. Some key takeaways are as follows:

- VQA is essential for ensuring viewer satisfaction by evaluating the video quality across various platforms.

- Some quality of experience (QoE) metrics include availability, bitrate, playback failure, rebuffering startup time, video start error, and lag ratio.

- Some quality of service (QoS) metrics include traffic/hits, throughput, offload, and availability.

- Some examples of VQA software currently available are Witbe and ViCue Soft.

- The incorporation of both subjective and objective indicators in VQA poses one of the primary challenges.

- The integration of machine learning algorithms is revolutionizing VQA, enabling more accurate and context-aware assessments of video quality.

In the next chapter, the machine learning techniques surrounding both NTMA and VQA used to build the applications in this book are mathematically broken down.

3.5 References – Chapter 3

1.   K. Bouraqia, E. Sabir, M. Sadik, and L. Ladid, "Quality of Experience for Streaming Services: Measurements, Challenges and Insights," *IEEE Access*, vol. 8, pp. 13341–61, 2020, doi: https://doi.org/10.1109/access.2020.2965099.

2.   T. P. Fowdur, B. N. Baulum, and Y. Beeharry, "Performance Analysis of Network
     Traffic Capture Tools and Machine Learning Algorithms for the Classification
     of Applications, States and Anomalies," *International Journal of Information
     Technology*, vol. 12, no. 3, pp. 805–24, Apr. 2020, doi: https://doi.org/10.1007/
     s41870-020-00458-0.

3.   A. Biernacki and K. Tutschku, "Performance of HTTP Video Streaming Under
     Different Network Conditions," *Multimedia Tools and Applications*, vol. 72, no. 2,
     pp. 1143–66, Mar. 2013, doi: https://doi.org/10.1007/s11042-013-1424-x.

4.   F. Loh, F. Poignée, F. Wamser, F. Leidinger, and T. Hoßfeld, "Uplink vs. Downlink:
     Machine Learning-Based Quality Prediction for HTTP Adaptive Video Streaming,"
     Sensors, vol. 21, no. 12, p. 4172, Jun. 2021, doi: https://doi.org/10.3390/s21124172.

5.   MantisNet, "Network Traffic Analysis: Real-time Identification, Detection and
     Response to Threats," www.mantisnet.com, https://www.mantisnet.com/blog/
     network-traffic-analysis (accessed Jan. 22, 2024).

6.   E. Zerman, Baris Konuk, G. Nur, and Gözde Bozdağı Akar, "A Parametric Video
     Quality Model Based on Source and Network Characteristics," 2014 IEEE
     International Conference on Image Processing (ICIP), Oct. 2014, doi: https://doi.
     org/10.1109/icip.2014.7025119.

7.   J. Korhonen, "Two-Level Approach for No-Reference Consumer Video Quality
     Assessment," *IEEE Transactions on Image Processing*, vol. 28, no. 12, pp. 5923–38,
     Dec. 2019, doi: https://doi.org/10.1109/tip.2019.2923051.

8.   S. Mustafa and A. Hameed, "Perceptual Quality Assessment of Video Using
     Machine Learning Algorithm," Signal, Image and Video Processing, May 2019, doi:
     https://doi.org/10.1007/s11760-019-01494-5.

9.   J. Søgaard, Søren Forchhammer, and J. Korhonen, "Video Quality Assessment
     and Machine Learning: Performance and Interpretability," Technical University
     of Denmark, DTU Orbit (Technical University of Denmark, DTU), May 2015, doi:
     https://doi.org/10.1109/qomex.2015.7148149.

10. "Papers with Code - Video Quality Assessment," paperswithcode.com, https://
     paperswithcode.com/task/video-quality-assessment (accessed Jan. 22, 2024).

11. International Telecommunication Union, "Telephone Transmission Quality,
     Telephone Installations, Local Line Networks," www.itu.int, https://www.itu.
     int/rec/T-REC-P/en (accessed Jan. 22, 2024).

12. International Telecommunication Union, "BT.500: Methodologies for the Subjective
     Assessment of the Quality of Television Images," www.itu.int, https://www.itu.
     int/rec/R-REC-BT.500/en.

13. Kalpana Seshadrinathan, R. Soundararajan, A. C. Bovik, and L. K. Cormack, "A Subjective Study to Evaluate Video Quality Assessment Algorithms," Proceedings of SPIE, Feb. 2010, doi: https://doi.org/10.1117/12.845382.

14. M. Muniz, "Video Quality Metrics," InTech eBooks, Feb. 2010, doi: https://doi.org/10.5772/8038.

15. Ann Marie Rohaly et al., "Video Quality Experts Group: Current Results and Future Directions," Proceedings of SPIE, May 2000, doi: https://doi.org/10.1117/12.386632.

16. International Telecommunication Union, "P.910: Subjective Video Quality Assessment Methods for Multimedia Applications," www.itu.int, Oct. 2023, https://www.itu.int/rec/T-REC-P.910.

17. International Telecommunication Union, "P.913: Methods for the Subjective Assessment of Video Quality, Audio Quality and Audiovisual Quality of Internet Video and Distribution Quality Television in Any Environment," www.itu.int, Jun. 2021, https://www.itu.int/rec/T-REC-P.913.

18. International Telecommunication Union, "P.1204: Video Quality Assessment of Streaming Services Over Reliable Transport for Resolutions up to 4K," www.itu.int, Oct. 2023, https://www.itu.int/rec/T-REC-P.1204 (accessed Jan. 22, 2024).

19. X. Min, K. Gu, L. Zhang, Vinit Jakhetiya, and G. Zhai, "Editorial: Computational Neuroscience for Perceptual Quality Assessment," Frontiers in Neuroscience, vol. 16, Mar. 2022, doi: https://doi.org/10.3389/fnins.2022.876969.

20. A. van Kasteren, K. Brunnström, J. Hedlund, and C. Snijders, "Quality of Experience of 360 Video – Subjective and Eye-Tracking Assessment of Encoding and Freezing Distortions," Multimedia Tools and Applications, vol. 81, no. 7, pp. 9771–802, Feb. 2022, doi: https://doi.org/10.1007/s11042-022-12065-1.

21. T. Sun, S. Ding, and W. Chen, "Blind Video Quality Assessment Based on Multilevel Video Perception," Signal Processing: Image Communication, vol. 99, p. 116485, Nov. 2021, doi: https://doi.org/10.1016/j.image.2021.116485.

22. M. Majidi, "Video Advertising and Marketing Worldwide," Statista, Dec. 18, 2023, https://www.statista.com/topics/5960/digital-video-advertising/#topicOverview (accessed Jan. 22, 2024).

23. PricewaterhouseCoopers, "Streaming Ahead: Making UX and Content Strategy Work Together," PwC, 2019, https://www.pwc.com/us/en/services/consulting/library/consumer-intelligence-series/streaming-ahead.html.

24. Akamai, "Measuring Video Quality and Performance: Best Practices," May 2020, https://www.akamai.com/site/it/documents/white-paper/measuring-video-quality-and-performance-best-practices.pdf (accessed Jan. 22, 2024).

25.   P. Kafka, "Netflix Data: 70 Percent of Viewing Happens on TVs," *Vox*, Mar. 07, 2018. `https://www.vox.com/2018/3/7/17094610/netflix-70-percent-tv-viewing-statistics` (accessed Jan. 22, 2024).

26.   Akamai, "Increase Viewer Loyalty: Best Practices for Ensuring a Quality OTT Experience," Sep. 2018, `https://www.akamai.com/site/en/documents/white-paper/increase-viewer-loyalty-through-high-quality-ott-video-streaming-whitepaper.pdf` (accessed Jan. 22, 2024).

27.   Akamai, "What Does 'Good' Look Like?" Sep. 2018, `https://www.akamai.com/site/fr/documents/white-paper/what-does-good-look-like-ott-video-quality.pdf` (accessed Jan. 22, 2024).

28.   Consumer Technology Association, "Streaming Quality of Experience Events, Properties and Metrics (CTA-2066)," Consumer Technology Association, Mar. 2020, `https://shop.cta.tech/products/streaming-quality-of-experience-events-properties-and-metrics` (accessed Jan. 22, 2024).

29.   Tricentis, "Top 5 Metrics for Streaming Video Performance," Tricentis, Aug. 03, 2021, `https://www.tricentis.com/blog/top-5-metrics-for-streaming-video-performance` (accessed Jan. 22, 2024).

30.   S. Pham, Cise Midoglu, R. Seeliger, S. Arbanowski, and S. Steglich, "A Novel Approach to Streaming QoE Score Calculation by Integrating Error Impacts," SOICT '23: Proceedings of the 12th International Symposium on Information and Communication Technology, Dec. 2023, doi: https://doi.org/10.1145/3628797.3628985.

31.   A. Wishnu and B. Sugiantoro, "Analysis of Quality Of Service (QoS) YouTube Streaming Video Service in Wireless Network in the Environment Faculty of Science and Technology Uin Sunan Kalijaga," *IJID (International Journal on Informatics for Development)*, vol. 7, no. 2, p. 30, Jan. 2019, doi: https://doi.org/10.14421/ijid.2018.07206.

32.   M. H. Pinson, L. Janowski, and Z. Papir, "Video Quality Assessment: Subjective Testing of Entertainment Scenes," *IEEE Signal Processing Magazine*, vol. 32, no. 1, pp. 101–14, Jan. 2015, doi: https://doi.org/10.1109/msp.2013.2292535.

33.   E. Technology, "Video Quality Assessment," Medium, Apr. 12, 2018, `https://eyevinntechnology.medium.com/video-quality-assessment-34abd35f96c0` (accessed Jan. 22, 2024).

34.   H. R. Wu and K. R. Rao, *Digital Video Image Quality and Perceptual Coding*, CRC Press, 2017.

35.   H. Fu, D. Pan, and P. Shi, "Full-Reference Video Quality Assessment Based on Spatiotemporal Visual Sensitivity," 2021 International Conference on Culture-oriented Science & Technology (ICCST), Nov. 2021, doi: https://doi.org/10.1109/iccst53801.2021.00071.

36.   Z. Wang and A. Bovik, "Reduced- and No-Reference Image Quality Assessment," *IEEE Signal Processing Magazine*, vol. 28, no. 6, pp. 29–40, Nov. 2011, doi: https://doi.org/10.1109/msp.2011.942471.

37.   X. Zhang, L. Xie, and Z. Guo, "Quality Assessment and Measurement Quality Assessment and Measurement for Internet Video Streaming for Internet Video Streaming," *ZTE Communications*, vol. 17, no. 1, Mar. 2019, doi: https://doi.org/10.12142/ZTECOM.201901003.

38.   Z. Sinno and A. C. Bovik, "Large-Scale Study of Perceptual Video Quality," *IEEE Transactions on Image Processing*, vol. 28, no. 2, pp. 612–27, Feb. 2019, doi: https://doi.org/10.1109/tip.2018.2869673.

39.   N. Ponomarenko et al., "Image Database TID2013: Peculiarities, Results and Perspectives," *Signal Processing: Image Communication*, vol. 30, pp. 57–77, Jan. 2015, doi: https://doi.org/10.1016/j.image.2014.10.009.

40.   Y. Wang, S. Inguva, and B. Adsumilli, "YouTube UGC Dataset for Video Compression Research," *IEEE Xplore*, pp. 1–5, Sep. 2019, doi: https://doi.org/10.1109/MMSP.2019.8901772.

41.   N. T. Blog, "Toward a Practical Perceptual Video Quality Metric," Medium, Apr. 19, 2017, `https://netflixtechblog.com/toward-a-practical-perceptual-video-quality-metric-653f208b9652` (accessed Jan. 22, 2024).

42.   Video Processing, Compression and Quality Research Group, "MSU Video Quality Metrics Benchmark Methodology," videoprocessing.ai, Mar. 12, 2022, `https://videoprocessing.ai/benchmarks/video-quality-metrics_methodology.html` (accessed Jan. 22, 2024).

43.   Manish Narwaria and L. Wang, "Video Quality Assessment Using Temporal Quality Variations and Machine Learning," 2011 IEEE International Conference on Multimedia and Expo, Jul. 2011, doi: https://doi.org/10.1109/icme.2011.6011936.

44.   X. Gao, G. Liu, W. Lu, D. Tao, and X. Li, "Spatio-Temporal Salience Based Video Quality Assessment," 2010 IEEE International Conference on Systems, Man and Cybernetics, Oct. 2010, doi: https://doi.org/10.1109/icsmc.2010.5642429.

45. Z. Tu, C.-J. Chen, Y. Wang, N. Birkbeck, Balu Adsumilli, and A. C. Bovik, "Video Quality Assessment of User Generated Content: A Benchmark Study and a New Model," 2021 IEEE International Conference on Image Processing (ICIP), Sep. 2021, doi: https://doi.org/10.1109/icip42928.2021.9506189.

46. Z. Tu, Y. Wang, N. Birkbeck, B. Adsumilli, and A. C. Bovik, "UGC-VQA: Benchmarking Blind Video Quality Assessment for User Generated Content," *IEEE Transactions on Image Processing*, vol. 30, pp. 4449–64, 2021, doi: https://doi.org/10.1109/tip.2021.3072221.

47. Y. Liu, J. Wu, L. Li, W. Dong, and G. Shi, "Quality Assessment of UGC Videos Based on Decomposition and Recomposition," *IEEE Transactions on Circuits and Systems for Video Technology*, vol. 33, no. 3, pp. 1043–54, Mar. 2023, doi: https://doi.org/10.1109/tcsvt.2022.3209007.

48. J. Ruan and D. Xie, "A Survey on QoE-Oriented VR Video Streaming: Some Research Issues and Challenges," *Electronics*, vol. 10, no. 17, p. 2155, Sep. 2021, doi: https://doi.org/10.3390/electronics10172155.

49. IPTV, "What Are the Emerging Trends and Technologies That Will Impact IPTV Video Quality Assessment in the Future?" LinkedIn, https://www.linkedin.com/advice/3/what-emerging-trends-technologies-impact-iptv-video-quality (accessed Jan. 22, 2024).

50. W. Moina-Rivera, J. Aguado, and M. García-Pineda, "Video Quality Metrics Toolkit: An Open Source Software to Assess Video Quality," *SoftwareX*, vol. 23, Jul. 2023, doi: https://doi.org/10.1016/j.softx.2023.101427.

51. Telestream, "Video Test and Monitoring Equipment, Sentry Software-Based Quality Monitoring Solution | Telestream," www.telestream.net, 2024, https://www.telestream.net/iq/sentry.htm (accessed Jan. 22, 2024).

52. ThinkTel, "Real Time Media Assessment for Voice and Video Network Analysis," ThinkTel, 2024, https://www.thinktel.ca/services/real-time-media-assessment/ (accessed Jan. 22, 2024).

53. "Witbe," www.witbe.net, 2024. https://www.witbe.net/ (accessed Jan. 22, 2024).

54. Witbe, "Quality of Experience Witbe and Standards about Witbe," 2008 (accessed: Jan. 22, 2024) [Online], https://www.csimagazine.com/pdf/witbe_and_standards_en.pdf.

55. Witbe, "Google's Stadia Cloud Gaming Quality Analysis," www.witbe.net, https://www.witbe.net/resources/google-cloud-gaming-quality-analysis/ (accessed Jan. 22, 2024).

56. "Witbe Unveils VQ-ID - Connected Media|IP," www.connectedmedia-ip.com, https://www.connectedmedia-ip.com/news/witbe/witbe-unveils-vq-id (accessed Jan. 22, 2024).

57. Witbe, "KEY FEATURES" (accessed Jan. 22, 2024) [Online], https://vectorsolutions.net/wp-content/uploads/import/Witbe%20TT%20Robots.pdf.

58. ViCue Soft, "VQ Analyzer User Guide," vicuesoft.com, 2024, https://vicuesoft.com/vq-analyzer/userguide/ (accessed Jan. 22, 2024).

59. AccepTV, "Video Quality Monitor," www.acceptv.com, 2024, https://www.acceptv.com/en/products_vqm.php (accessed Jan. 22, 2024).

60. AccepTV, "No Reference, Parametric and Hybrid Perceived Video Quality Measurement/Monitoring Solution" (accessed Jan. 22, 2024) [Online], https://www.acceptv.com/en/media/documents/AccepTV%20-%20Video%20Quality%20Monitor%20-%20Datasheet.pdf.

61. Video Quality Experts Group (VQEG), "Video Quality Experts Group (VQEG)," www.vqeg.org, ehttps://www.vqeg.org/vqeg-home/ (accessed Jan. 22, 2024).

62. Video Quality Experts Group (VQEG), "A Brief Introduction to the Video Quality Experts Group," Jan. 2023, https://www.vqeg.org/media/cyrkpdwb/vqeg_introduction_2022.pdf (accessed Jan. 22, 2024).

63. rkalidin, "Releases · VIQET/VIQET-Desktop," GitHub, Mar. 19, 2016, https://github.com/VIQET/VIQET-Desktop/releases (accessed Jan. 22, 2024).

64. Wyzowl, "Video Marketing Statistics 2023," Wyzowl, 2023, https://www.wyzowl.com/video-marketing-statistics/.

65. M. Solomons, "200 Key Video Marketing Statistics and Insights," Linearity blog, Jul. 02, 2023, https://www.linearity.io/blog/video-marketing-statistics/ (accessed Jan. 22, 2024).

66. L. Forristal, "Netflix Now Lets TV Viewers Customize Text for Subtitles and Closed Captions," Yahoo Finance, Mar. 08, 2023, https://uk.finance.yahoo.com/news/netflix-now-lets-tv-viewers-212627725.html?guccounter=2 (accessed Jan. 22, 2024).

67. D. Li, T. Jiang, and J. Ming, "Recent Advances and Challenges in Video Quality Assessment," *ZTE Communications*, vol. 17, no. 1, pp. 3–11, Nov. 2019, doi: https://doi.org/10.12142/ztecom.201901002.

68. C. Aditya, "The Power of Quality Video in Advertising," www.linkedin.com, Jun. 15, 2021, https://www.linkedin.com/pulse/power-quality-video-advertising-candace-aditya-/ (accessed Jan. 22, 2024).

69. Abrar Majeedi, B. Naderi, Yasaman Hosseinkashi, J. Cho, R. Martínez, and R. Cutler, "Full Reference Video Quality Assessment for Machine Learning-Based Video Codecs," arXiv (Cornell University), Sep. 2023, doi: https://doi.org/10.48550/arxiv.2309.00769.

70. P. V. Vu and D. M. Chandler, "ViS3: An Algorithm for Video Quality Assessment Via Analysis of Spatial and Spatiotemporal Slices," *Journal of Electronic Imaging*, vol. 23, no. 1, p. 013016, Feb. 2014, doi: https://doi.org/10.1117/1.jei.23.1.013016.

71. S. Li, F. Zhang, L. Ma, and K. N. Ngan, "Image Quality Assessment by Separately Evaluating Detail Losses and Additive Impairments," *IEEE Transactions on Multimedia*, vol. 13, no. 5, pp. 935–49, Oct. 2011, doi: https://doi.org/10.1109/tmm.2011.2152382.

72. M. Xu, J. Chen, H. Wang, S. Liu, G. Li, and Z. Bai, "C3DVQA: Full-Reference Video Quality Assessment with 3D Convolutional Neural Network," arXiv (Cornell University), Oct. 2019, doi: https://doi.org/10.48550/arxiv.1910.13646.

73. W. Sun, X. Min, W. Lu, and G. Zhai, "A Deep Learning–based No-reference Quality Assessment Model for UGC Videos," Proceedings of the 30th ACM International Conference on Multimedia, Oct. 2022, doi: https://doi.org/10.1145/3503161.3548329.

74. D. C. Mocanu, A. Liotta, A. Ricci, M. T. Vega, and G. Exarchakos, "When Does Lower Bitrate Give Higher Quality in Modern Video Services?" 2014 IEEE Network Operations and Management Symposium (NOMS), May 2014, doi: https://doi.org/10.1109/noms.2014.6838400.

75. T. Hobfeld, R. Schatz, M. Varela, and C. Timmerer, "Challenges of QoE Management for Cloud Applications," *IEEE Communications Magazine*, vol. 50, no. 4, pp. 28–36, Apr. 2012, doi: https://doi.org/10.1109/mcom.2012.6178831.

76. M. Torres Vega, C. Perra, F. De Turck, and A. Liotta, "A Review of Predictive Quality of Experience Management in Video Streaming Services," *IEEE Transactions on Broadcasting*, vol. 64, no. 2, pp. 432–45, Jun. 2018, doi: https://doi.org/10.1109/tbc.2018.2822869.

77. S. Mohamed, G. Rubino, F. Cervantes, and H. Afifi, "Real-Time Video Quality Assessment in Packet Networks: A Neural Network Model," `Inria.fr`, 2022, doi: `https://hal.inria.fr/inria-00072437`.

78. P. M. Ruiz, J. A. Botia, and A. Gomez-Skarmeta, "Providing QoS Through Machine-Learning-Driven Adaptive Multimedia Applications," *IEEE Transactions on Systems, Man and Cybernetics, Part B (Cybernetics)*, vol. 34, no. 3, pp. 1398–411, Jun. 2004, doi: https://doi.org/10.1109/tsmcb.2004.825912.

79. J. Jagadessan, B. Nikita, G. D. Preta, and H. H. Priya, "A Machine Learning Algorithm for Jitter Reduction and Video Quality Enhancement in IoT Environment," *International Journal of Engineering and Advanced Technology (IJEAT)*, vol. 8, no. 4, Apr. 2019.

80. N. Thomos, T. Maugey, and L. Toni, "Machine Learning for Multimedia Communications," *Sensors*, vol. 22, no. 3, p. 819, Jan. 2022, doi: https://doi.org/10.3390/s22030819.

81. F. Li, Y. Zhang, and P. C. Cosman, "MMMNet: An End-to-End Multi-Task Deep Convolution Neural Network with Multi-Scale and Multi-Hierarchy Fusion for Blind Image Quality Assessment," *IEEE Transactions on Circuits and Systems for Video Technology*, vol. 31, no. 12, pp. 4798–811, Dec. 2021, doi: https://doi.org/10.1109/tcsvt.2021.3055197.

82. J. Fu, X. Chen, Z. Zhang, S. Wu, and Z. Chen, "360SRL: A Sequential Reinforcement Learning Approach for ABR Tile-Based 360 Video Streaming," 2019 IEEE International Conference on Multimedia and Expo (ICME), Jul. 2019, doi: https://doi.org/10.1109/icme.2019.00058.

83. N. Kan, J. Zou, K. Tang, C. Li, N. Liu, and H. Xiong, "Deep Reinforcement Learning-based Rate Adaptation for Adaptive 360-Degree Video Streaming," ICASSP 2019 - 2019 IEEE International Conference on Acoustics, Speech and Signal Processing (ICASSP), May 2019, doi: https://doi.org/10.1109/icassp.2019.8683779.

84. M. Torres Vega, D. C. Mocanu, and A. Liotta, "Unsupervised Deep Learning for Real-Time Assessment of Video Streaming Services," *Multimedia Tools and Applications*, vol. 76, no. 21, pp. 22303–27, May 2017, doi: https://doi.org/10.1007/s11042-017-4831-6.

85. Baris Konuk, E. Zerman, G. Nur, and Gözde Bozdağı Akar, "Video Content Analysis Method for Audiovisual Quality Assessment," 2016 Eighth International Conference on Quality of Multimedia Experience (QoMEX), Jun. 2016, doi: https://doi.org/10.1109/qomex.2016.7498965.

86. D. Raca et al., "On Leveraging Machine and Deep Learning for Throughput Prediction in Cellular Networks: Design, Performance, and Challenges," *IEEE Communications Magazine*, vol. 58, no. 3, pp. 11–17, Mar. 2020, doi: https://doi.org/10.1109/mcom.001.1900394.

87. H. Mao, R. Netravali, and M. Alizadeh, "Neural Adaptive Video Streaming with Pensieve," Proceedings of the Conference of the ACM Special Interest Group on Data Communication, Aug. 2017, doi: https://doi.org/10.1145/3098822.3098843.

88. H. Yousef, J. L. Feuvre, and A. Storelli, "ABR Prediction Using Supervised Learning Algorithms," 2020 IEEE 22nd International Workshop on Multimedia Signal Processing (MMSP), Sep. 2020, doi: https://doi.org/10.1109/mmsp48831.2020.9287123.

89. A. Bentaleb, C. Timmerer, A. C. Begen, and R. Zimmermann, "Bandwidth Prediction in Low-Latency Chunked Streaming," Proceedings of the 29th ACM Workshop on Network and Operating Systems Support for Digital Audio and Video, Jun. 2019, doi: https://doi.org/10.1145/3304112.3325611.

90. Y. Zhao, Q.-W. Shen, W. Li, T. Xu, W.-H. Niu, and S.-R. Xu, "Latency Aware Adaptive Video Streaming using Ensemble Deep Reinforcement Learning," Proceedings of the 27th ACM International Conference on Multimedia, Oct. 2019, doi: https://doi.org/10.1145/3343031.3356071.

91. Y. Sani, D. Raca, J. J. Quinlan, and C. J. Sreenan, "SMASH: A Supervised Machine Learning Approach to Adaptive Video Streaming over HTTP," 2020 Twelfth International Conference on Quality of Multimedia Experience (QoMEX), May 2020, doi: https://doi.org/10.1109/qomex48832.2020.9123139.

92. D. da Hora, K. van Doorselaer, K. van Oost, and R. Teixeira, "Predicting the Effect of Home Wi-Fi Quality on QoE," inria.hal.science, Jan. 06, 2018, https://inria.hal.science/hal-01676921 (accessed May 14, 2023).

93. A. Ligata, E. Perenda, and H. Gacanin, "Quality of Experience Inference for Video Services in Home WiFi Networks," IEEE Communications Magazine, vol. 56, no. 3, pp. 187–93, Mar. 2018, doi: https://doi.org/10.1109/mcom.2018.1700712.

CHAPTER 4

Machine Learning Techniques for NTMA and VQA

This chapter delves into the machine learning (ML) techniques that surround and enable network traffic monitoring and analysis (NTMA) and video quality assessment (VQA). Through passive listening of network parameters' being reported by the network interface, the Node.js server formulates a series of arrays that keep track of network traffic collected over time. The same applies to ML-derived mean opinion score (MOS) values through video streaming. Techniques such as multilayer perceptron (MLP) are used for regression—i.e., the prediction of both network traffic and video QoS scores—while the classification used is the K-nearest neighbors (KNN) model. This chapter contains a thorough mathematical breakdown of all the algorithms coded in this book for both classification and prediction of network traffic metrics, together with the production of an MOS score for video quality.

4.1 Classification Model for NTMA

In this book, the two types of ML approaches to network traffic are classification and prediction, where the former is carried out using the K-nearest neighbor (KNN) model. This algorithm is elaborated upon in this section, where it allows the end user to obtain a real-time activity status of the device on the network, among three broad categories.

© Tulsi Pawan Fowdur, Lavesh Babooram 2024
T. P. Fowdur, L. Babooram, *Machine Learning For Network Traffic and Video Quality Analysis*,
https://doi.org/10.1007/979-8-8688-0354-3_4

4.1.1 Data Collection for Classification

The first stage in classification includes pre-categorizing the data being gathered, as the goal is to predict outcomes that typically belong to different groups or categories. Specifically, different user activities such as "idle," "browsing," and "streaming" are determined based on real-time network traffic. As a result, parameters inside these predefined network contexts are captured to build the training dataset. In this case, the recorded values are classified according to the most probable type of traffic using the following states:

- **Idle**: The data-collecting device is not actively sending and receiving any network-related metrics to or from its network interfaces, with the exception of some peaks observed through notifications or messages from social media applications. Otherwise, the network characteristics remain rather stable and around zero.

- **Browsing**: This includes web surfing and scrolling through social media applications like Facebook and Google News, where network traffic is often observed to be of an erratic nature with frequent spikes. Browsing can be considered moderate usage.

- **Streaming**: The device is used to stream videos on platforms such as Twitch or YouTube, while network traffic is collected. This usually results in high bandwidth usage depending on the platform and quality of the video being watched.

A portion of such a network traffic classification dataset in a 3D space, including another parameter—namely, "Packets Received"—is shown in Figure 4-1.

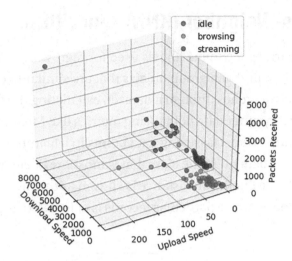

Figure 4-1. *3D representation of device activity states*

At one-second intervals, the network's operational status is used to manually assign a label to each row, before being kept locally on the monitoring device. Examples of some values in this dataset are given next, where the last digit denotes the category assigned, namely 0, 1, and 2, which point to the three device activities:

- **Idle:** 1, 1, 0

- **Idle:** 1, 2, 0

- **Browsing:** 210, 31, 1

- **Browsing:** 462, 44, 1

- **Streaming:** 1049, 20, 2

- **Streaming:** 1190, 21, 2

Such datasets can be built around a larger number of device states, such as gaming, video conferencing, social media usage, and online shopping. A sample of the classification dataset is provided along with the book codes. It is, however, recommended that each user performs their own data collection tailored to the specifications of their device and internet connectivity package, such that an optimum accuracy is achieved during classification.

4.1.2 K-Nearest Neighbor (KNN) Algorithm

KNN is a supervised learning technique that is used for regression and classification problems. Its purpose is to find and group objects that are similar to each other [1]. This approach iteratively searches through a dataset, calculating the distance between the test instance and every other data point using metrics like Minkowski, Manhattan, or the Euclidean distance. When classifying data points in a feature space, it uses a voting method to give the data point the class that is most prevalent among its k-closest neighbors. In this book, the Euclidean distance is chosen as the measurement technique, where the distance d between two particular records is calculated as given in Equation 4-1 [2]:

$$d(p,q) = \sqrt{\sum_{i=1}^{n}(q_i^2 - p_i^2)} \qquad (4\text{-}1)$$

Where:

- p, q: Two points in Euclidean n-space

- qi, pi: Euclidean vectors, starting from the origin of space

- n: Number of axes making the plane

4.1.3 Data Preparation for Classification

Before feeding data into the KNN model, it needs to be processed into a suitable format. A snippet of the three activity states as the training data and a sample test data is shown in Figure 4-2. At this point, a large dataset requires shuffling before being input to the model.

	Download Speed	Upload Speed	Pre-Assigned Category
Training Data	2	11	0
	129	25	1
	1856	58	2
Test Data	1276	47	?

Figure 4-2. *Data preparation for classification*

4.1.4 Shorthand Example for KNN

The following steps pertain to a numerical example to perform KNN classification for network data based on the model coded in this book.

1. Consider the set of values, together with their pre-labeled classes given in Table 4-1. The set of metrics to be classified is also shown.

Table 4-1. Sample Data for KNN Classification

Download Speed	Upload Speed	Device Activity	Numerical class
2	11	Idle	0
129	25	Browsing	1
1856	58	Streaming	2
1276	47	undefined	undefined

2. Thus, the Euclidean distance for one comparison can be calculated as follows:

$$x_1 = \sqrt{(1276-2)^2 + (47-11)^2} = 1274.508533$$

3. This is repeated for all the records present in the database where the "undefined" record in Table 4-1 is compared against all records such that a logical table is built containing the distances for each record comparison.

4. Table 4-2 thus shows the Euclidean distance from the record being classified to every other set of values. This is mainly performed to cast the number of votes.

Table 4-2. *Euclidean Distances Calculated*

Euclidean Distance	Class
1274.508533	0
1147.210966	1
580.104301	2

5. The number of neighbors defined, however, has a say in what class is allocated to the record. For example, if $k = 1$ in this case, it means that the k closest neighbor's class is selected. From Table 4-2, the record is thus classified as "streaming" with the closest class being "2," given the shortest Euclidean distance.

4.2 Prediction Model for NTMA

In this section, the multilayer perceptron (MLP) regression model is described. It is used in conjunction with the sliding or rolling window method to create tuples of data fed into the model for training. Together with a hyperparameter optimization process, the MLP model is used to create a time-series forecasting framework, allowing the user to use the "Window Size" and "Prediction Time" variables to predict network traffic for any timestamp.

4.2.1 Multilayer Perceptron (MLP) Algorithm

The MLP algorithm belongs to the neural network realm and uses the backpropagation approach for the training process [3]. This includes forward and backward passes where parameters known as weights are adjusted such that they are tailored for the particular dataset. The weights are initially assigned at random, before the model aims to decrease the error between the predicted value and the data point input. This creates a continuous loop where the error keeps decreasing until specific requirements set by the user are met [4]. MLP is a popular model for classification and regression problems [5], as it starts with an input layer that takes in pre-processed data, followed by a certain number of hidden layers that feed-forward computing units. The output layer is then connected to these hidden layers to generate the prediction. A basic representation of an MLP model with two hidden layers is given in Figure 4-3.

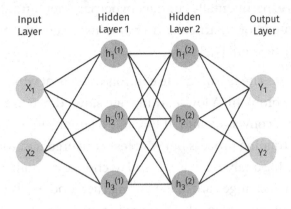

Figure 4-3. *MLP architecture*

A simple MLP can be summarized by Equation 4-2 [6]:

$$o^0 = x, \; o^l = F^l\left(W^l \underline{o}^{(l-1)}\right) \; for \; l = 1,\dots,L \tag{4-2}$$

Where:

- x: Input vector, set as the "output of the zeroth layer"

- l: Layer in question

- o^l: Output vector

- $\underline{o}^{(l-1)}$: Increasing the dimension of a vector by prepending a number 1

- W^l: Bias term of a particular layer

- F^l: Activation function applied to all components of a parameter

4.2.2 Hyperparameters

The MLP model is usually governed by hyperparameters that dictate the network structure and ultimately set the tone for how the network is trained. They are set before the training process. The ideal values can differ according to the distinct attributes of the dataset, including its magnitude, intricacy, and data distribution. Given that network traffic data is dependent on underlying factors such as network bandwidth, networking devices, and the device being used to collect data, the hyperparameters that yield the best performance accuracy may vary from user to user and scenario to scenario. It is thus recommended to experimentally conduct hyperparameter tuning to determine the best combination of values for a particular user or network environment. Some typical hyperparameters are listed here [7, 8]:

- **Learning Rate**: The learning rate controls the rate at which the model adapts to the problem. A low learning rate slows down the learning procedure but converges smoothly, while a larger value may result in a quicker training process but with less convergence. To achieve accuracy during analytics, it is advised to choose a small learning rate such that small changes are made during each update, but this is at the expense of more time taken for prediction.

- **Epochs**: This is the total number of iterations/passes of all the training data in one cycle during the training process.

- **Number of Hidden Layers**: A satisfactory number of hidden layers results in highly accurate values, together with less time complexity. However, a high number of layers causes the neural network to become more intricate [9].

4.2.3 Data Preparation for Time-Series Prediction

MLP regression requires data to be normalized before being input into the neural network. This usually causes the training process to speed up, thus reducing convergence time [10]. The min-max normalization method is used to fit the data between 0 and 1, as per Equation 4-3 [11]:

$$X_{normalized} = \frac{X - X_{min}}{X_{max} - X_{min}}$$

(4-3)

Where:

- X: Parameter undergoing normalization

- X_{max}: Maximum value of X

- X_{min}: Minimum value of X

Likewise, after prediction, the output needs to be denormalized, such that the value is upscaled by the same ratio. Equation 4-4 is then implemented:

$$X_{denormalized} = X_{normalized}\left(X_{max} - X_{min}\right) + X_{min} \qquad (4\text{-}4)$$

Next, the principle of sliding windows is applied to the normalized dataset before feeding each window into the algorithm for training. The pre-processing of a univariate model fed into an MLP neural network is adapted from Neuroph's neural network documentation [12, 13, 14]. This is described in the next section.

4.2.4 Sliding Window Concept

Also known as the rolling window, the sliding window concept is efficient for short time-series forecasting models [15, 16]. This is ideal for the regression methods in this book since the application takes as input a particular prediction time defined by the user. Consider the scenario given next, which consists of "Download Speed" values taken from a dataset with the following parameters for the regression model:

- **Window Size**: 5 seconds

- **Prediction Time**: 2 seconds

For explanatory purposes, the values given in Figure 4-4 are obtained directly from the dataset. These are normalized before constructing the sliding windows, as shown next.

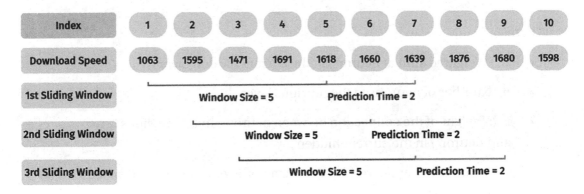

Figure 4-4. *Example of sliding windows for download speed*

This means that for the first sliding window, the first five instances are taken as the input data, and the (window size + prediction time)th value, in this case the seventh value, is taken as the output data. The index is incremented by one, and the loop continues until the end of the dataset. This concept trains the MLP model to understand that it must use the last sliding window to forecast the desired value, dictated by the prediction time.

4.2.5 MLP for Time-Series Network Traffic Prediction

In this book, the sliding window method is used, together with separate data collections, as dictated by the window size and prediction time factors entered by the end user. A univariate approach is thus considered during the data processing phase for each of the three variables. Equations 4-5 to 4-7 are used to calculate the weighted sum at each neuron in the hidden layers for download speed, upload speed, and latency:

$$z_{j^h} = \sum_{1}^{m} \left(w_{ij^h} D_i \right) + b_{j^h} \tag{4-5}$$

$$z_{j^h} = \sum_{1}^{m} \left(w_{ij^h} U_i \right) + b_{j^h} \tag{4-6}$$

$$z_{j^h} = \sum_{1}^{m} \left(w_{ij^h} L_i \right) + b_{j^h} \tag{4-7}$$

Where:

- z_{j^h}: Weighted sum at neuron j in the current hidden layer h

- m: Number of neurons in the previous layer

- w_{ij^h}: Weight of the connection between neuron i in the previous layer and neuron j in the current hidden layer

- D_i: Input values at neuron i in the previous layer for download speed

- U_i: Input values at neuron i in the previous layer for upload speed

- L_i: Input values at neuron i in the previous layer for latency

- *i*: Time step of each sliding window, which also represents the actual neurons in the previous layer

- b_{jh}: Random bias representation for neuron *j* in the current hidden layer

Equation 4-8 illustrates the calculation of the weighted sum at the output neuron after applying the activation function F^l. This equation represents the prediction output for the parameter:

$$z_{1^{output}} = \sum_{1}^{n} \left(w_{ij^{output}} F^l_{j((2 \cdot window\ size)+1)} \right) + b_{1^{output}}$$
(4-8)

Where:

- $z_{1^{output}}$: Weighted sum at the output neuron

- 1: Denotes the single output layer

- *n*: Number of neurons in the last hidden layer

4.2.6 Short-hand Example for MLP

In the example shown in Table 4-3, a univariate approach for the prediction of latency values is broken down using MLP. The aim is to use the first two values to predict the latency for the third timestep. This thus denotes a typical example of the training process.

Table 4-3. *Example of Prediction of Latency Values using MLP*

Timestep	Latency (ms)
1	29.2
2	35.8
3	31.0
4	37.0
5	26.8
6	25.0

The steps presented next are followed:

1. Given that the maximum value in the dataset is 37.0, with the minimum being 25.0, the values are normalized according to Equation 4-3. This gives Table 4-4.

Table 4-4. *Normalization Process for Latency Values*

Timestep	Latency	Normalized Value
1	29.2	0.35
2	35.8	0.90
3	31.0	0.50
4	37.0	1.00
5	26.8	0.15
6	25.0	0.00

2. Since only the first three values are being considered, 0.35 and 0.90 are used as the input values, with 0.50 as the expected output value. Figure 4-5 shows the neural network that is constructed using the first two normalized values with randomly assigned weights.

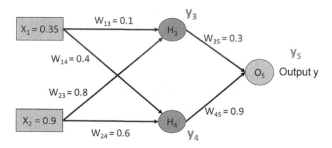

Figure 4-5. *Neural network for latency prediction*

3. In this example, a forward pass is followed by a backward pass, and another forward pass is performed. The actual output is assumed to be 0.5 and the learning rate to be 1. Equation 4-8 gives the Sigmoid activation function, and Equation 4-9, the forward pass formula:

$$y_j = f(a_j) = \frac{1}{1 + e^{-a_j}} \tag{4-8}$$

$$a_j = \sum_j (w_{i,j} * x_j) \tag{4-9}$$

4. For the forward pass, the outputs for Y_3, Y_4, and Y_5 are computed:

    a. At H_3,
        $a_1 = (W_{13} * X_1) + (W_{23} * X_2)$
        $= (0.1 * 0.35) + (0.8 * 0.9)$
        $= 0.755$
        Thus, $Y_3 = f(a_1) = f(1 / (1 + e^{-0.755}))$
        $= 0.6802671967$

    b. At H_3,
        $a_1 = (W_{13} * X_1) + (W_{23} * X_2)$
        $= (0.1 * 0.35) + (0.8 * 0.9)$
        $= 0.755$
        Thus, $Y_4 = f(a_2) = f(1 / (1 + e^{-0.68}))$
        $= 0.6637386974$

    c. At O_5,
        $a_3 = (W_{35} * Y_3) + (W_{45} * Y_4)$
        $= (0.3 * 0.6802671967) + (0.9 * 0.6637386974)$
        $= 0.8014449867$
        Thus, $Y_5 = f(a_3) = f(1 / (1 + e^{-0.8014449867}))$
        $= 0.6902834929$

The network output for the first forward pass is therefore 0.6902834929.

5. Since the actual target output value is 0.5, the error is computed using Equation 4-10.

$$Error = Y_{target} - Y_5 \qquad (4\text{-}10)$$

Therefore,

$$Error = 0.5 - 0.6902834929$$

$$Error = -0.19$$

6. This error can now be used to verify if the requirement of the model is satisfied or not. The weights are updated through the backpropagation method to reduce this error. To determine by how much each weight has changed, the following equations are used:

$$\Delta w_{ji} = \eta \delta_j o_i \qquad (4\text{-}11)$$

$$\delta_j = o_j \left(1 - o_j\right)\left(t_j - o_j\right) \textit{ if } j \textit{ is an output unit} \qquad (4\text{-}12)$$

$$\delta_j = o_j \left(1 - o_j\right)\sum_k \delta_k w_{kj} \textit{ if } j \textit{ is a hidden unit} \qquad (4\text{-}13)$$

Where:

- η: Is a constant called the learning rate

- t_j: Is the correct teacher output for unit j

- δ_j: Is the error measure for unit j

This means that δ_j is required at all units.

a. For the output unit,
$\delta_5 = Y_5 \left(1 - Y_5\right) \left(Y_{target} - Y_5\right)$
$= 0.6903 \left(1 - 0.6903\right) \left(0.5 - 0.6903\right)$
$= -0.04068112511$

b. For hidden units H_3,

$\delta_3 = Y_3 (1 - Y_3) (\delta_5 * W_{35})$

$= 0.6803 (1 - 0.6803) (-0.04068112511 * 0.3)$

$= -0.00265448903$

c. For hidden units H_4,

$\delta_4 = Y_4 (1 - Y_4) (\delta_5 * W_{45})$

$= 0.6637 (1 - 0.6637) (-0.04068112511 * 0.9)$

$= -0.008171645064$

7. The backward pass now consists of calculating the final Δw term at every unit, which is the modified weight. For example, Δw_{45} is taken as follows:

$\Delta W_{45} = n\, \delta_j\, O_i$

$= n\, \delta_5\, Y_4$

$= (1) (-0.04068112511) (0.6637)$

$= -0.02700163699$

Therefore,

$\Delta w_{45\,(new)} = \Delta w_{45} + \Delta w_{45\,(old)}$

$= -0.02700163699 + 0.9$

$= 0.872998363$

This is repeated for all other weights, yielding Table 4-5.

Table 4-5. *Weights Update Process*

i	j	w_{ij}	δ_j	x_i	n	Updated w_{ij}
1	3	0.1	-0.00265	0.35	1	0.0991
2	3	0.8	-0.00265	0.9	1	0.7976
1	4	0.4	-0.0082	0.35	1	0.3971
2	4	0.6	-0.0082	0.9	1	0.5926
3	5	0.3	-0.0406	0.6803	1	0.2724
4	5	0.9	-0.0406	0.6637	1	0.8731

8. Now, a forward pass is again computed for Y_3, Y_4, and Y_5:

    a. At H_3,

    $$a_1 = (W_{13} * X_1) + (W_{23} * X_2)$$
    $$= (0.0991 * 0.35) + (0.7976 * 0.9)$$
    $$= 0.7525$$

    Thus, $Y_3 = f(a_1) = f(1 / (1 + e^{-0.7525}))$
    $$= 0.6797$$

    b. At H_4,

    $$a_2 = (W_{14} * X_1) + (W_{24} * X_2)$$
    $$= (0.3971 * 0.35) + (0.5926 * 0.9)$$
    $$= 0.6723$$

    Thus, $Y_4 = f(a_2) = f(1 / (1 + e^{-0.6723}))$
    $$= 0.6620$$

    c. At O_5,

    $$a_3 = (W_{35} * Y_3) + (W_{45} * Y_4)$$
    $$= (0.2724 * 0.6797) + (0.8731 * 0.6620)$$
    $$= 0.7631$$

    Thus, $Y_5 = f(a_3) = f(1 / (1 + e^{-0.7631}))$
    $$= 0.6820$$

    The network output for the second forward pass is therefore 0.6820.

9. With the weights updated and the network output calculated, the error is computed as follows:

    $$Error = 0.5 - 0.6820$$

    $$Error = -0.182$$

    This means that the error has been reduced from -0.19 to -0.182. At this point, the maximum error rate hyperparameter decides whether the model performs more passes.

10. Finally, the network output is denormalized using Equation 4-4, thus giving the following result:

$$X_{denormalized} = \left[X_{normalised} \left(X_{max} - X_{min} \right) \right] + X_{min}$$

$$X_{denormalized} = \left[0.6820 \left(37 - 25 \right) \right] + 25$$

$$X_{denormalized} = 33.184$$

This means that the predicted value is 33.184 ms.

11. The percentage accuracy for this specific prediction can be calculated as shown here:

$$\% Accuracy = \frac{X - X_{minimum}}{X_{maximum} - X_{minimum}} * 100$$

$$\% Accuracy = \frac{33.184 - 31.0}{31.0} * 100$$

$$Accuracy = 7.04\%$$

4.3 SVM for VQA

In this book, the live objective mean opinion score (MOS) is generated through continuous screenshots being fed into a hosted Java servlet, where a no-reference (NR) method is used for image quality assessment (IQA) using the Blind/Reference-less Image Spatial Quality Evaluator (BRISQUE) principles [17, 18]. A new screenshot is taken and processed as soon as the video quality score for the previous one is obtained, resulting in a continuous loop.

The degree of distortion in an image is therefore determined through artifacts such as blocking, ringing, blurring, and noising by applying the series of steps detailed in this section. For this method, a smaller score indicates better video quality. This metric is denoted as an objective MOS given the widespread use and acknowledgment among VQA vendors in the industry [19]. This section thus breaks down how the support vector machine (SVM) algorithm is used for regression as the ML algorithm in producing the video quality score from a live video, which is pre-processed as an image. As shown in the work outlined by Min et al. [20], the VQA block in this book uses the blind image quality assessment technique using distortion aggravation.

4.3.1 Blind Image Quality Assessment Using Distortion Aggravation

The methodology presented in this book for VQA can be divided into three main steps, as follows [20]:

1. Different kinds of degradation aggravations are introduced during the generation of the multiple pseudo reference images (MPRIs) to estimate the most probable type of distortion in the image. The MPRIs are produced by adding different ratios of the following degradations [21]:

    a. JPEG Compression

    b. JPEG2000 Compression

    c. Gaussian Blur

    d. White Noise

2. The local binary pattern (LBP) features are then extracted from each image before assessing the similarity between the image and each of the generated MPRIs [22].

3. The similarity scores are fed into an SVM regression model for scaling and final forecast of the video quality score [23].

4.3.2 Preliminary Steps

Figure 4-6 shows the system model around which the BRISQUE method is based.

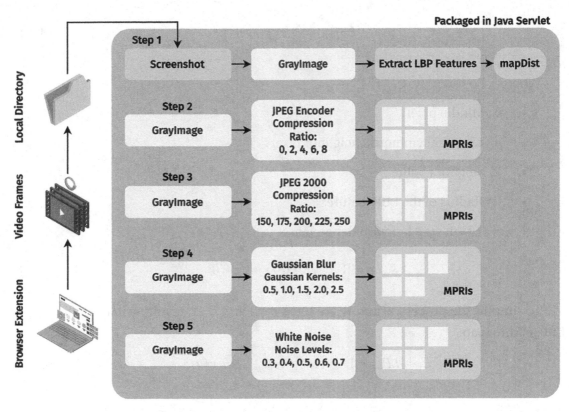

Figure 4-6. *Preliminary steps for the BRISQUE method*

The Java servlet is responsible for receiving a "number" tag denoting which screenshot it should fetch from the local directory specified by the user within the codes. Once received, it reads the cropped screenshot before converting it to grayscale, as indicated by the GrayImage object in Figure 4-6.

4.3.3 Extraction of LBP Features

In this section, the extraction of LBP features is explained with reference to both mathematical equations and illustrations as performed in code.

Process and Equations

When the grayscale image is obtained, LBP features are extracted from the grayImage object, with the results stored in the mapDist array. The binarization process is conducted with the subsequent luminance contrasts as per Equation 4-14 [22]:

$$LBP_{V,W} = \sum_{v=0}^{V-1} z(p_n - p_c) \tag{4-14}$$

Where:

- p_c: Middle pixel

- p_n: Circular symmetric neighbor pixel

- V: Neighbor value

- W: Radius of the LBP feature

- $z(*)$: Unit step function denoted as

$$z = \{1 \qquad z \geq 0 \, 0 \qquad z < 0 \tag{4-15}$$

To simplify the process, V and W are set as $p = 4$ and $r = 1$, resulting in the following LBP feature map:

$$LP_{pr} = \{1 \qquad LP_{4,1} = c \, 0 \qquad otherwise \tag{4-16}$$

Where:

- p and r: Pixel indices

- c: Indicates the five LBPs for $0 \leq c \leq 5$

The LBP extraction process is first performed for the original image, with the same method performed for the degraded image and the MPRIs. This results in feature maps labeled LP_d and LP_m.

Illustrations and Coding Procedure

This process is performed using the LBP41() method. As an example, the matrix of pixels given in Figure 4-7 is considered where its values are stored in the img1[][] array.

287	178	63	78	255	75	84	278	222	173	65
90	243	91	110	84	208	146	250	149	9	158
187	167	201	32	187	118	217	93	17	250	200
251	245	11	108	79	130	58	232	20	71	194
101	99	123	218	85	119	178	250	141	248	120
47	68	130	21	178	255	123	14	250	104	250
178	98	123	250	12	225	172	107	87	252	155

Figure 4-7. *Grayscale image pixel values*

The steps listed next are then taken:

1. The center pixels are extracted from img1[ ][ ] and stored in the center[ ][ ] array, as demonstrated in Figure 4-8 by the shaded area.

287	178	63	78	255	75	84	278	222	173	65
90	243	91	110	84	208	146	250	149	9	158
187	167	201	32	187	118	217	93	17	250	200
251	245	11	108	79	130	58	232	20	71	194
101	99	123	218	85	119	178	250	141	248	120
47	68	130	21	178	255	123	14	250	104	250
178	98	123	250	12	225	172	107	87	252	155

Figure 4-8. *Extraction of center pixel values*

2. Then, the pixels from img1[ ][ ] are extracted at four different neighboring offsets before being stored in the 3D cell[ ][ ][ ] array, as shown in the next steps.

    a. An offset of (0, 1) is first considered, as shown in Figure 4-9.

287	178	63	78	255	75	84	278	222	173	65
90	243	91	110	84	208	146	250	149	9	158
187	167	201	32	187	118	217	93	17	250	200
251	245	11	108	79	130	58	232	20	71	194
101	99	123	218	85	119	178	250	141	248	120
47	68	130	21	178	255	123	14	250	104	250
178	98	123	250	12	225	172	107	87	252	155

Figure 4-9. *Extraction of pixel values at offset (0, 1)*

b. This is followed by an offset of (-1, 0), as depicted in Figure 4-10.

287	178	63	78	255	75	84	278	222	173	65
90	243	91	110	84	208	146	250	149	9	158
187	167	201	32	187	118	217	93	17	250	200
251	245	11	108	79	130	58	232	20	71	194
101	99	123	218	85	119	178	250	141	248	120
47	68	130	21	178	255	123	14	250	104	250
178	98	123	250	12	225	172	107	87	252	155

Figure 4-10. *Extraction of pixel values at offset (-1, 0)*

c. Likewise, Figure 4-11 shows the extraction at an offset of (0, 1).

287	178	63	78	255	75	84	278	222	173	65
90	243	91	110	84	208	146	250	149	9	158
187	167	201	32	187	118	217	93	17	250	200
251	245	11	108	79	130	58	232	20	71	194
101	99	123	218	85	119	178	250	141	248	120
47	68	130	21	178	255	123	14	250	104	250
178	98	123	250	12	225	172	107	87	252	155

Figure 4-11. *Extraction of pixel values at offset (0,-1)*

d. Finally, an offset of (1, 0) is taken, as shown in Figure 4-12.

287	178	63	78	255	75	84	278	222	173	65
90	243	91	110	84	208	146	250	149	9	158
187	167	201	32	187	118	217	93	17	250	200
251	245	11	108	79	130	58	232	20	71	194
101	99	123	218	85	119	178	250	141	248	120
47	68	130	21	178	255	123	14	250	104	250
178	98	123	250	12	225	172	107	87	252	155

Figure 4-12. *Extraction of pixel values at offset (1,0)*

3. Next, each pixel from the center matrix is compared with each pixel from each of the four neighboring matrices. If the pixel value from the cell[][][] array is greater than that of the center[][] array, the resulting pixel value in cell[][][] is 1. Otherwise, it is 0. This is illustrated in Figure 4-13.

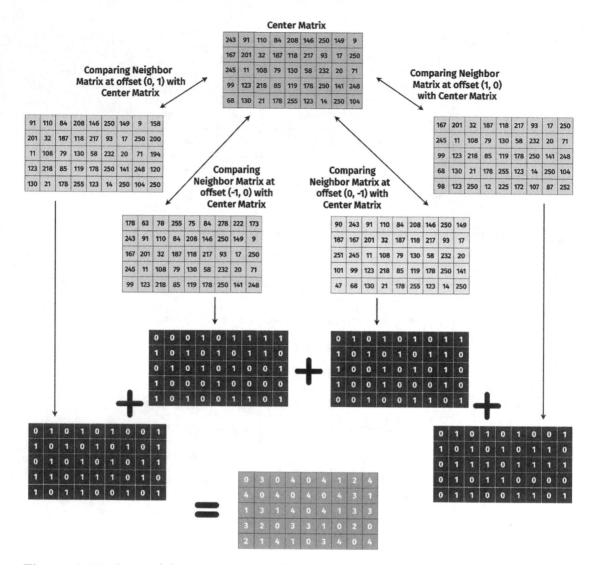

Figure 4-13. *Sum of the comparison of center and neighbor matrices*

4. Finally, each of the LBP[][] matrices is binarized according
    to its distortion type in the LBP41jpeg(), LBP41jpeg2k(),
    LBP41gblur(), and LBP41wnoise() methods. This is shown in
    Figure 4-14.

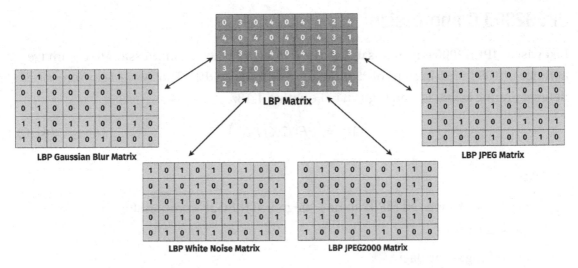

Figure 4-14. *Binarization process based on the corresponding distortion type*

4.3.4 Distortion Aggravation

The grayscale image is subject to four different types of compressions and distortions to generate the MPRIs.

JPEG Compression

The image is compressed at compression ratios ranging from 0 to 8 in steps of 2. The type of artifact for which the image is being tested is blocking. At the end of this process, a set of five images known as the MPRIs are output, as shown in Figure 4-6. JPEG compression is outlined by Equation 4-17, where *I* denotes the original image.

$$M_{lj} = JPEG(I, C_j) \tag{4-17}$$

Where:

- M_{lj}: Compressed MPRI at the jth degree of JPEG compression for $0 \le j \le 5$

- l: Blocking artifact

- *JPEG*: JPEG encoder

- C_j: Compression ratio

JPEG2000 Compression

Likewise, a JPEG2000 encoder is used to introduce JPEG2000 compression to the image at ratios of 150 to 250 in steps of 25, thus generating its MPRIs. This type of compression detects ringing artifacts, and is given in Equation 4-18:

$$M_{rj} = JPEG2K\left(I, N_j\right)$$ (4-18)

Where:

- M_{rj}: Compressed MPRI at the jth degree of JPEG2K compression for $0 \leq j \leq 5$

- r: Ringing artifact

- $JPEG2K$: JPEG2K encoder

- N_j: Compression ratio

Gaussian Blur

Similarly, Gaussian blur is added to the grayscale image with degradation levels ranging from 0.5 to 2.5 in steps of 0.5 to detect the blurring artifact. This is given in Equation 4-19:

$$M_{bj} = GK_j * I$$ (4-19)

Where:

- M_{bj}: Blurred MPRI at the jth degree of blur for $0 \leq j \leq 5$

- b: Blurring artifact

- GK_j: Gaussian kernel

White Noise

Finally, the last distortion type added is white noise at levels ranging from 0.3 to 0.7 in steps of 0.1. This detects noising and is given by Equation 4-20:

$$M_{nj} = I + N\left(0, V_j\right)$$ (4-20)

Where:

- M_{nj}: Noisy MPRI at the jth degree of white noise for $0 \leq j \leq 5$

- n: Noising artifact

- $N(0, V_j)$: Produces normally distributed random values with mean 0 and variance V_j

4.3.5 Similarity Index

After applying the different types of compressions and distortions, a set of twenty MPRIs is generated. LBP features are again extracted from each of them before storing the resulting arrays in an array denoted as mapPRI. This is illustrated in Figure 4-15 [23, 24].

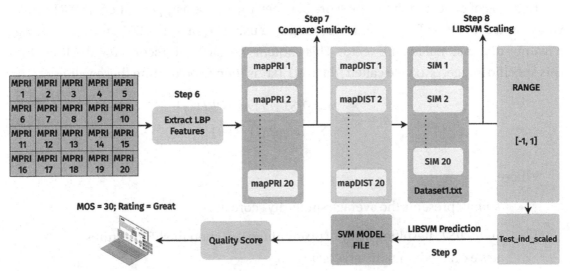

Figure 4-15. *Similarity index and video quality score prediction model*

The overlap between the LP_d and LP_m feature maps is calculated using Equation 4-21:

$$LP_o = \left(LP_d . LP_m \right) \tag{4-21}$$

Where:

- LP_o: Overlap between LP_d and LP_m

- LP_d: Feature map of the degraded image

- LP_m: Feature map of the MPRIs

The similarity, *sim*, is defined by Equation 4-22, where the numerator and denominator provide an indication of the number of non-zero elements in the feature maps:

$$sim = s\left(LP_d, LP_m\right) = \frac{\sum_{p,r} LPo_{p,r}}{\sum_{p,r} LPm_{p,r}} \tag{4-22}$$

In terms of code, the similarity score between the previously generated `mapPRI` array values and the `mapDIST` array values is computed using Equation 4-23, in turn producing a twenty-dimensional feature vector. This contains the similarity scores for all MPRIs for this specific image. A .txt file called "dataset1.txt" is then used to store these values.

$$sim = \frac{\sum_{i,j} mapDIST\left(i,j\right).mapPRI\left(i,j\right)}{\sum_{i,j} mapPRI\left(i,j\right)+1} \tag{4-23}$$

Where:

- *sim*: Represents the average similarity score

- *mapDIST*: Resulting array of pixels obtained when the LBP features are extracted from the grayscale image

- *mapPRI*: Array of pixels obtained when the LBP features are extracted from each of the MPRIs

- *i* and *j*: Indices of the values in the *mapDIST* and *mapPRI* arrays

The twenty-dimensional feature vector notation is given in Equation 4-24:

$$sim = \left[sim_{l1},..,sim_{l5},\ sim_{r1},..,sim_{r5},\ sim_{b1},..,sim_{b5}, sim_{n1},..,sim_{n5} \right] \tag{4-24}$$

Where:

- sim_{l1} to sim_{l5}: Similarity scores of feature vectors subject to blocking

- sim_{r1} to sim_{r5}: Similarity scores of feature vectors subject to ringing

- sim_{b1} to sim_{b5}: Similarity scores of feature vectors subject to blurring

- sim_{n1} to sim_{n5}: Similarity scores of feature vectors subject to noising

4.3.6 Scaling

For this step, the "range" file is used as it contains the minimum and maximum scaling factors for each index, where the former are listed below -1, and the latter, below 1. The piece of code for the scaling process uses the following files:

- "dataset1.txt": Contains the test data

- "range": Contains the values used to scale the twenty-dimensional feature vector to [-1.0, +1.0]

- "test_ind_scaled": Contains the scaled values

The "range" file is shown in Figure 4-16.

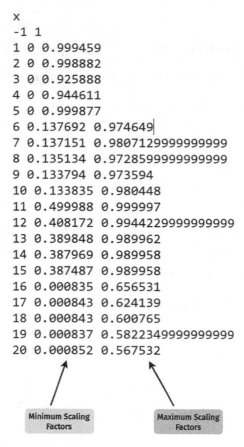

```
x
-1 1
1 0 0.999459
2 0 0.998882
3 0 0.925888
4 0 0.944611
5 0 0.999877
6 0.137692 0.974649
7 0.137151 0.9807129999999999
8 0.135134 0.9728599999999999
9 0.133794 0.973594
10 0.133835 0.980448
11 0.499988 0.999997
12 0.408172 0.9944229999999999
13 0.389848 0.989962
14 0.387969 0.989958
15 0.387487 0.989958
16 0.000835 0.656531
17 0.000843 0.624139
18 0.000843 0.600765
19 0.000837 0.5822349999999999
20 0.000852 0.567532
```

Minimum Scaling Factors

Maximum Scaling Factors

Figure 4-16. *Scaling factors in the range file*

The minimum and maximum scaling factors are taken from this file and used as the lower and upper limits, respectively, for scaling the feature vector from the "dataset1.txt" file. Equation 4-25 summarizes the scaling process:

$$scale_{ind} = -1.0 + (1.0 + 1.0) \times \left(\frac{value_{ind} - feat\_min_{ind}}{feat\_max_{ind} - feat\_min_{ind}} \right) \qquad (4\text{-}25)$$

Where:

- $scale_{ind}$: Scaled-value-at-index-ind

- $value_{ind}$: Test value from the feature vector at index ind

- $feat\_min_{ind}$: Minimum-scaling-factor-at-index-ind

- $feat\_max_{ind}$: Maximum-scaling-factor-at-index-ind

Each calculated value is then printed into the "test_ind_scaled" file.

4.3.7 Using SVM for Prediction

To predict the overall image quality score from the similarity scores obtained, the twenty-dimensional vector is fed into an SVM for regression. The LIBSVM library [25] is used based on its simplicity, efficiency, and reliability in solving forecasting problems. The "MOS" score is thus calculated as per Equation 4-26 using the svm_predict() method:

$$FIN = svm\_predict(\mathrm{mod}, regressor) \text{ for } j \in d \qquad (4\text{-}26)$$

Where:

- *FIN*: Final video quality score (MOS)

- *regressor*: Calculated-scaled-features

- *mod*: Model around which prediction-is-approximated

The piece of code for the prediction method uses the following files:

- "test_ind_scaled": Contains the scaled values, thus the test data for prediction

- "model": Contains the dataset model

- "output.txt": Contains the output results

Therefore, the svm_predict() method approximates the test data according to the parameters from the "model" file. A quality rating is also given along with the video quality score. Once determined, the servlet writes the score back to the client in a JSON format.

4.4 Summary

Chapter 4 consisted of the mathematical concepts for the NTMA and VQA models built in this book. The NTMA classification model was created using the KNN algorithm, while the prediction model used the MLP neural network with fine-tuned hyperparameters for optimal accuracy and performance. A short-hand example has been given for both frameworks, describing the concepts and equations used for generating the device activity and prediction value at any time instant. Likewise, for VQA, the SVM algorithm was used for scaling and prediction of the MOS score. Some key takeaways are as follows:

- Machine learning concepts are of significant importance in augmenting the functionalities of both NTMA and VQA frameworks.

- In the context of NTMA, the classification model allows the device activity to be determined among three states—namely, idle, browsing, and streaming—based on live network traffic classified against a pre-labeled dataset.

- Prediction algorithms facilitate the forecast of network parameters for NTMA and the calculation of an MOS score for VQA.

- The BRISQUE method for determining live video quality is thoroughly described.

The next chapter consists of the NTMA application development, including the system model, client and server structures, and a testing and deployment section.

4.5 References – Chapter 4

1. K. Bouraqia, E. Sabir, M. Sadik, and L. Ladid, "Quality of Experience for Streaming Services: Measurements, Challenges and Insights," *IEEE Access*, vol. 8, pp. 13341–61, 2020, doi: https://doi.org/10.1109/access.2020.2965099.

2. C. Sammut and G. Webb, *Encyclopedia of Machine Learning*. New York, Springer, 2011, p. 903.

3. C. M. Bishop, *Neural Networks for Pattern Recognition*. Oxford: Oxford University Press, 2002.

4. P. Vadlamudi, "Machine Learning: Using Machine Learning Algorithms to Provide Local Temperature Prediction," doi: https://doi.org/10.13140/RG.2.2.21007.79523.

5. S. Abirami and P. Chitra, "Energy-efficient Edge-based Real-time Healthcare Support System," *Advances in Computers*, vol. 117, no. 1, pp. 339–68, 2020, doi: https://doi.org/10.1016/bs.adcom.2019.09.007.

6.   P. Nieminen, "Multilayer Perceptron Training with Multiobjective Memetic Optimization," *Jyväskylä studies in computing*, no. 247, 2016 (accessed Feb. 18, 2024) [Online], `http://urn.fi/URN:ISBN:978-951-39-6824-3`.

7.   T. Elansari, M. Ouanan, and H. Bourray, "Modeling of Multilayer Perceptron Neural Network Hyperparameter Optimization and Training," Feb. 2023, doi: `https://doi.org/10.21203/rs.3.rs-2570112/v1`.

8.   Rendyk, "Tuning the Hyperparameters and Layers of Neural Network Deep Learning," Analytics Vidhya, May 26, 2021. `https://www.analyticsvidhya.com/blog/2021/05/tuning-the-hyperparameters-and-layers-of-neural-network-deep-learning/` (accessed Feb. 18, 2024).

9.   M. Uzair and N. Jamil, "Effects of Hidden Layers on the Efficiency of Neural networks," *IEEE Xplore*, Nov. 1, 2020, `https://ieeexplore.ieee.org/document/9318195`.

10.   T. Stöttner, "Why Data Should Be Normalized Before Training a Neural Network," Medium, May 16, 2019, `https://towardsdatascience.com/why-data-should-be-normalized-before-training-a-neural-network-c626b7f66c7d` (accessed Feb. 18, 2024).

11.   S. Loukas, "Everything You Need to Know About Min-Max Normalization in Python," Medium, Jun. 14, 2020, `https://towardsdatascience.com/everything-you-need-to-know-about-min-max-normalization-in-python-b79592732b79` (accessed Feb. 18, 2024).

12.   V. Steinhauer, "Prices Prediction with Feed-Forward Neural Networks," `neuroph.sourceforge.net`, `http://neuroph.sourceforge.net/tutorials/ChickenPricePredictionTutorial.htm` (accessed Feb. 18, 2024).

13.   V. Steinhauer, "Stock Market Prediction with Feed-Forward Neural Networks," `neuroph.sourceforge.net`, `http://neuroph.sourceforge.net/tutorials/StockMarketPredictionTutorial.html` (accessed Feb. 18, 2024).

14.   L. E. Carter-Greaves, "Time Series Prediction with Feed-Forward Neural Networks," `neuroph.sourceforge.net`, `http://neuroph.sourceforge.net/TimeSeriesPredictionTutorial.html` (accessed Feb. 18, 2024).

15.   D. Overload, "Sliding Window Technique — Reduce the Complexity of Your Algorithm," Medium, Dec. 21, 2022, `https://medium.com/@data-overload/sliding-window-technique-reduce-the-complexity-of-your-algorithm-5badb2cf432f` (accessed Feb. 18, 2024).

16. J. Brownlee, "Time Series Forecasting as Supervised Learning," Machine Learning Mastery, Dec. 04, 2016, `https://machinelearningmastery.com/time-series-forecasting-supervised-learning/` (accessed Feb. 18, 2024).

17. K. R. Shrimali, "Image Quality Assessment: BRISQUE | Learn OpenCV," Jun. 20, 2018, `https://learnopencv.com/image-quality-assessment-brisque/` (accessed Feb. 18, 2024).

18. MATLAB, "Blind/Referenceless Image Spatial Quality Evaluator (BRISQUE) No-Reference Image Quality Score - MATLAB BRISQUE," MathWorks, `https://www.mathworks.com/help/images/ref/brisque.html` (accessed Feb. 18, 2024).

19. R. C. Streijl, S. Winkler, and D. S. Hands, "Mean Opinion Score (MOS) Revisited: Methods and Applications, Limitations and Alternatives," *Multimedia Systems*, vol. 22, no. 2, pp. 213–27, Dec. 2014, doi: `https://doi.org/10.1007/s00530-014-0446-1`.

20. X. Min, G. Zhai, K. Gu, Y. Liu, and X. Yang, "Blind Image Quality Estimation via Distortion Aggravation," *IEEE Transactions on Broadcasting*, vol. 64, no. 2, pp. 508–17, Jun. 2018, doi: `https://doi.org/10.1109/TBC.2018.2816783`.

21. X. Min, K. Gu, G. Zhai, J. Liu, X. Yang, and C. W. Chen, "Blind Quality Assessment Based on Pseudo-Reference Image," *IEEE Transactions on Multimedia*, vol. 20, no. 8, pp. 2049–62, Aug. 2018, doi: `https://doi.org/10.1109/tmm.2017.2788206`.

22. S. Allagwail, O. Gedik, and J. Rahebi, "Face Recognition with Symmetrical Face Training Samples Based on Local Binary Patterns and the Gabor Filter," *Symmetry*, vol. 11, no. 2, p. 157, Jan. 2019, doi: `https://doi.org/10.3390/sym11020157`.

23. M. Awad and R. Khanna, "Support Vector Regression," *Efficient Learning Machines*, pp. 67–80, 2015, doi: `https://doi.org/10.1007/978-1-4302-5990-9_4`.

24. Fowdur, T.P., Shaikh Abdoolla, M.A.N. and Doobur, L. (2023), "Performance Analysis of Edge, Fog and Cloud Computing Paradigms for Real-Time Video Quality Assessment and Phishing Detection," *International Journal of Pervasive Computing and Communications*, Vol. ahead-of-print, No. ahead-of-print, `https://doi.org/10.1108/IJPCC-09-2022-0327`

25. C.-C. Chang and C.-J. Lin, "LIBSVM — A Library for Support Vector Machines," `www.csie.ntu.edu.tw`, Jul. 09, 2023, `https://www.csie.ntu.edu.tw/~cjlin/libsvm/` (accessed Feb. 18, 2024).

CHAPTER 5

NTMA Application with JavaScript

The focus of this chapter shifts to application development based on the background elaborated in the previous chapters. First, a thorough system model will be presented. The architecture that encompasses the client–server interactions is detailed along with the functions and message of each block. In short, the browser extension coupled with the backend—i.e., Node.js—is built using a combination of HTML, CSS, and JavaScript to represent the client. With duplex communication between the client and the Node.js server, network traffic parameters are obtained from the network interface of the device before being subjected to prediction and classification Machine Learning (ML) techniques. As a result, parameters such as latency, jitter, and upload and download rates can be forecast in real time using a combination of the prediction time and window size variables set by the end user. This also applies for classification purposes where the real-time activity of the device is determined within a matter of seconds. Through thorough testing and deployment of the application in different networking scenarios, a functioning, reliable, and lightweight Network Traffic Monitoring and Analysis (NTMA) application is built.

5.1 System Model for NTMA

In this section, the system model for the NTMA framework is broken down into its constituent parts. The basic principle involves a client–server paradigm where the client is a browser extension and the server is a Node.js server running on either the same device or another one. The end user thus has a dashboard with the following capabilities:

© Tulsi Pawan Fowdur, Lavesh Babooram 2024
T. P. Fowdur, L. Babooram, *Machine Learning For Network Traffic and Video Quality Analysis*,
https://doi.org/10.1007/979-8-8688-0354-3_5

1. Monitor real-time network parameters directly in the web browser through a browser extension.

2. Obtain regression and classification analytics results based on historical data collated into real-time values.

3. Obtain a Auality of Service (QoS) score of the network performance.

The described functionalities are coded such that network traffic is monitored for a predefined and customizable period of time. The extension is coded with a blend of JavaScript (JS) and Node.js, where the latter provides server-side functionality and comprises the required libraries and access to the network interfaces for reading changes in network traffic. The frontend is made using a combination of HTML and CSS, while the extension and server communicate seamlessly in both directions using messages delivered and received in JSON format. These attributes all form part of the NTMA fabric, which is produced by the server upon request in a full duplex communication with the client, yielding both forecasting and classification using an array of ML techniques. After monitoring the different behaviors in the network trend, the backend triggers the analytics procedures, which act upon a combination of both historical data and live data, where the former is fetched from local databases stored on the device. This allows an increase in the performance accuracy of the algorithms during both classification and prediction. The most distinctive modern feature is the ability to use NTMA in a browser add-on through a remote server. Four essential files are needed for the entire extension package to push and retrieve data to and from the server, as shown in the complete system model depicted in Figure 5-1.

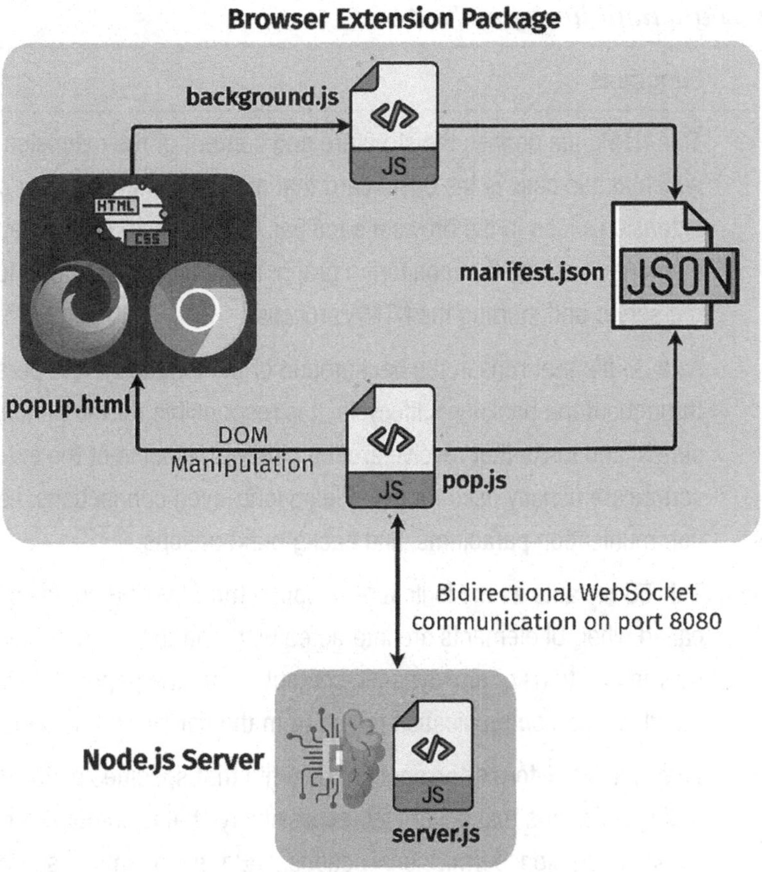

Browser Extension Package

Figure 5-1. *Complete NTMA system model*

5.1.1 Components and Functionalities

A constant two-way transmission of information in JSON format serves as the communication paradigm between the extension and the server. This is ensured by the WebSocket library [1], as it enables the delivery of tiny data chunks in both directions via a single, persistent connection. To send and receive messages to and from the server, the entire extension package needs four core files. These are summarized in Table 5-1.

Table 5-1. *File Functionalities for NTMA*

File	Functions
popup.html	This HTML file defines the structure and content of the extension's popup, which in this case is the dashboard that appears when the user clicks on the extension's icon in the browser's toolbar. It consists of the frontend, i.e., the UI elements, such as the monitoring graph, terminal, and a button for contacting the server and starting the NTMA process.
background.js	As a JS file that runs in the background of the extension and persists throughout the package's lifecycle, it is responsible for the handling of events, states, and tasks that revolve around different aspects of the extension. These scripts are usually used for overseeing long-lived connections, handling communication paradigms, and background actions.
pop.js	This JS file acts as the logic for the popup.html file and contains the functions called when UI elements are interacted with and triggered. It thus handles user interactions or any dynamic content within the popup. In this program, it handles data communication to and from the server and updates the DOM.
manifest.json	This is a JSON-formatted configuration file that specifies scripts to be used and permissions that govern tab accessibility. It also comprises metadata such as name, version, permissions, actions, and the icon image source for the extension. The browser obtains information about the extension from this file.
server.js	This JS file is run through Node.js and acts as the backend server for the extension. It handles communication to and from the client, i.e., the browser extension. In this application, it is responsible for monitoring network traffic parameters from the network interfaces and organizing them into time-series arrays, before performing prediction and classification, all while updating the monitoring graph at the client side in real time. Several libraries are housed in this file to allow this seamless interaction with the client.

When any online page is visited, the traffic extension can be used to observe and collect live network metrics for one minute, including download and upload speeds in Kbps and latency in ms. This is made feasible via the "systeminformation" Node. js package [2], which directly accesses the device's network interfaces, recording overall bandwidth involving both browser and background activities. The user enters a

"Prediction Time" parameter, which refers to the moment in time at which the variables will be forecast, as well as a "Window Size" field, which incorporates a principle usually used to construct ML models, as detailed previously in Chapter 4. The client site's dynamic graph is created with Chart JS [3].

5.1.2 Prediction and Classification of Network Traffic

In this study, the traffic flow gathered is passed through a series of data preparation steps before being fed into prediction and classification methods. This denotes the ML phase, where classification is performed using the K-nearest neighbors (KNN) algorithm, and prediction, using the multilayer perceptron (MLP) framework, which forms part of the neural network category. Since classification is concerned with the forecast of events that translate into classes, the first step is to pre-label the dataset. The aim here is to correctly interpret the network's traffic as "idle," "browsing," or "streaming." Consequently, the training dataset is built by collecting data in the aforementioned network settings. Every tuple is given a number label depending on the current state of the network's behavior at one-second intervals that is recorded. The dataset is then stored locally. To sum up, with an optimal number of five neighbors, the KNN algorithm is employed to dynamically determine the live state of the network, using the "ml-knn" Node module [4]. Likewise, the "mlp" node module [5] caters for MLP prediction to forecast the value of the passive network characteristics at the period of interest specified by the user. The Node.js server does all of the computations and then sends the results back to the client, where they are presented to the user.

5.1.3 NTMA Application Layout

For this application, a single window is built containing several components that facilitate the interaction of the user, with the network monitoring process occurring in the background. A live graph depicts the network traffic flow, which can be adjusted dynamically to view only the required parameter. The "Window Size" field contains a value that dictates the number of samples taken to forecast the predicted value, which is represented by the "Predict traffic for (s)" field, which in turn corresponds to the time at which the user wants the predicted values to occur. The predicted values for download and upload speeds, together with latency, are then given on the UI along with a calculated QoS score representing the network's performance in real time. These processes occur when the "Monitor & Analyze" button is pressed. Along with prediction capabilities, a description of the network activity state is also calculated through ML and given on the user interface, along with the average latency. The completed extension is as shown in Figure 5-2.

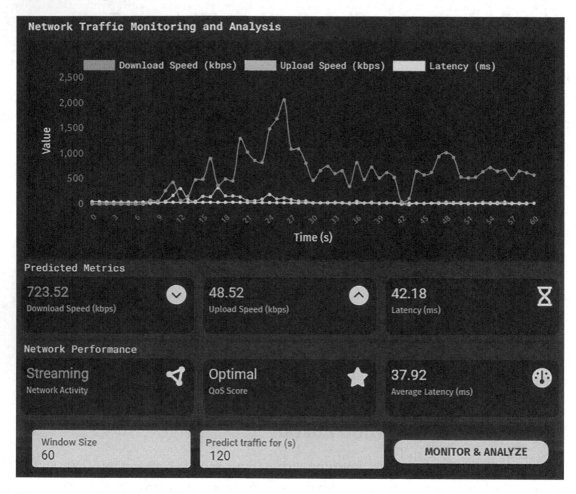

Figure 5-2. *NTMA application layout*

The parameters that are extracted, measured, and calculated are as follows:

- Download speed in kbps
- Upload speed in kbps
- Latency in ms
- Jitter
- Network QoS score

Figure 5-2 is a representation of live network monitoring, analytics, and a QoS score generated during a streaming scenario performed while watching a live video on Twitch. The score produced is "Optimal." With a prediction time of 120 seconds, the dashboard displays the predicted variables using MLP regression. The program also identifies automatically current device activity using KNN classification and past recorded network traffic states. The user may therefore monitor real-time network traffic characteristics directly from the browser, using a button that requests real-time classification and regression analytics from the Node.js server. The latter is locally hosted through a Node.js terminal, and each time the button is clicked, a new array of live parameters is collected for one minute.

5.1.4 Client–Server Interaction

Figure 5-3 depicts the client–server communication between the browser extension and the Node.js server in a stepwise manner.

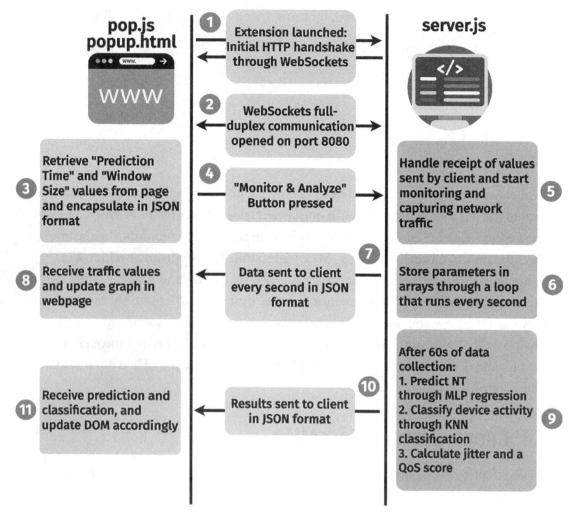

Figure 5-3. *Client–server interaction for NTMA*

5.2 Client Program Structure for NTMA

In this section, the four files making up the client side of the browser extension are described, together with their respective codes. An overview of the program structure for the client side is given in Figure 5-4.

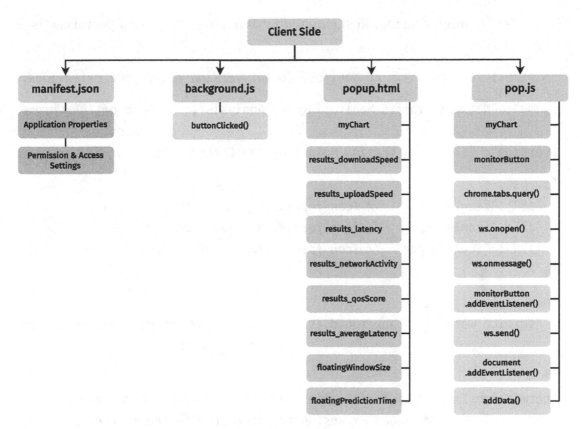

Figure 5-4. *Client side program structure*

5.2.1 Configuring Extension Settings and Permissions

As mentioned previously, the manifest.json file contains metadata that allows the browser to add it to the "Extensions" menu, and thus integrate it into Google Chrome. Listing 5-1 contains the parameters specified in this file, together with the respective purpose of each line in the comments. For the "icons" tag, it is recommended to add three icons of the user's choice with the naming conventions given in the code. These are PNG files that allow the user to easily identify the application on the browser extension menu. This tag can also be omitted in case the user prefers the default icon to appear instead.

Listing 5-1. Configuring manifest.json with extension settings and permissions

```
{
  "manifest_version": 3,  // Manifest format version for Chrome extensions.

  "name": "Network Traffic Monitoring and Analysis",  // Name of the Chrome
                                                          extension.
  "version": "1.0",  // Version number of the Chrome extension.

  "icons": {
    "16": "icon16.png",   // Icon for 16x16 pixel size.
    "48": "icon48.png",   // Icon for 48x48 pixel size.
    "128": "icon128.png"  // Icon for 128x128 pixel size.
  },

  "background": {
    "service_worker": "background.js"  // Background script, implemented as
                                          a service worker.
  },

  "permissions": [ // List of permissions required by the extension.
    "activeTab",  // Allows extension to interact with the currently
                     active tab.
    "scripting",  // Permits the injection of scripts into web pages.
    "webNavigation",  // Enables the extension to receive events in the web
                         navigation lifecycle.
    "webRequest",  // Grants the extension access to observe and analyze
                      network requests.
    "storage",  // Provides access to the browser's storage APIs for data
                   persistence.
    "declarativeNetRequest",  // Allows the extension to declare a set of
                                 rules for web requests.
    "declarativeNetRequestFeedback",  // Permits the extension to provide
                                         feedback on web requests.
    "tabs"  // Enables the extension to query and manipulate browser tabs.
  ],
```

```
"action": {
  "default_popup": "popup.html",  // Default popup HTML file when the
                                      extension icon is clicked.
  "default_title": "Network Traffic Monitoring and Analysis"
  // Default title for the extension icon.
}
}
```

5.2.2 Configuring the Background Script

As the silent orchestrator, the background script handles essential functionalities behind the scenes, such as button clicks, which then execute service worker logic. The lines of code required for this application are given in Listing 5-2 and are added to the background.js file.

Listing 5-2. Configuring background.js with service worker logic

```
// Log a message indicating that the extension has been launched.
console.log("Extension launched...");

// Listen for the extension's action button click event.
chrome.action.onClicked.addListener(buttonClicked);

// Function to handle the button click event.
function buttonClicked(tab) {
  // Log a message indicating the background script is running when the
  button is clicked.
  console.log("Background script running...");
}

// Additional service worker logic can be added here
// This script operates separately from the extension and can perform
background tasks efficiently
// Developers can extend the functionality of the service worker to handle
various tasks, such as
// network requests, data management, and background processing.
```

5.2.3 Building the User Interface

The frontend is built with a combination of HTML and CSS, where the latter is added to the same file as the HTML codes, thus following an internal CSS structure. The components in this section serve as the interface for network monitoring and analysis, with different sections such as the live graph, predicted metrics, network performance, and input fields for the end user.

File Functionality

In this section, a breakdown of the popup.html file is given, whose functions are as follows:

1. Specify the document structure, consisting of the following:

    i. Document type and language

    ii. Charset and title

    iii. Modify the content security policy (CSP) to allow certain script sources.

2. Add references to external resources:

    i. Import the required font stylesheets from Google Fonts [6].

    ii. Import icons for the dashboard from the Font Awesome icon library [7].

    iii. Link a custom Bootstrap stylesheet downloaded from Bootstrap's official website [8].

3. Add internal CSS for modifying the component appearance:

    i. Set styling for the body, background, and buttons.

    ii. Define styling for user interface components, such as input fields, labels, and charts.

4. Display a title and chart area:

    i. Give the popup a title.

    ii. Create a container for embedding the chart area.

    iii. Embed a chart canvas with Chart.js.

5. Display the "Predicted Metrics" cards:

   i. Create a container for predicted metrics.

   ii. Display the predicted download speed, upload speed, and latency individually along with their icons.

6. Display "Network Performance" cards:

   i. Create a container for network performance insights.

   ii. Display the network activity, QoS Score, and average latency individually along with their icons.

7. Add fields for user inputs:

   i. Define input fields for window size and prediction time.

   ii. Define the "Monitor & Analyze" button.

8. Add downloaded and client scripts:

   i. Link the downloaded Chart.js library [3].

   ii. Link the custom pop.js script.

Libraries and Required Resources

A summary of the mentioned libraries and downloadable scripts, together with their purpose, is given in Table 5-2.

Table 5-2. *Description of Libraries and Resources for the User Interface*

Library/ Resource	Purpose	Repository	Download Link	Location
Google Fonts	Provides custom fonts for the web page.	[6]		popup.html
Font Awesome	Provides scalable vector icons.	[7]		popup.html
bootstrap.min. css	Customized Bootstrap stylesheet for organizing row and columns.	[8]	[9]	"css" folder inside project's root folder
chart.js	Provides an interactive chart for the live graph.	[3]	[10]	Project's root folder
pop.js	Custom JS file for the logic.	N/A		Project's root folder

Creating the Document Structure

With the necessary libraries and resources at hand, the popup.html file can be populated with the codes in Listing 5-3, which starts off by creating the document structure. The placement of the codes for the rest of the file is indicated.

Listing 5-3. Creating the document structure

```html
<!DOCTYPE html>
<html lang="en" dir="ltr">
   <head>
      <!-- Listing 5-4 is included here -->
      <style>
      <!-- Listing 5-6 is included here -->
      </style>
   </head>
   <body>
      <!-- Listing 5-5 is included here -->
   </body>
</html>
```

Adding the Document Details and References to External Resources

Once the structure is set, the necessary fonts, icons, and styles are linked using the link tag and placed within the head tag, as given in Listing 5-4.

Listing 5-4. Adding document details and references to external resources

```html
<meta charset="utf-8">
<!-- Set Content Security Policy to allow scripts from the same origin
('self') and unsafe-eval for script execution -->
<meta http-equiv="Content-Security-Policy" content="script-src 'self'
'unsafe-eval'; object-src 'self'">
<!-- Title for the tab -->
<title>Network Monitoring and Analysis</title>
<!-- Font Stylesheet -->
```

```
<link rel="stylesheet" type="text/css" href="http://fonts.googleapis.com/cs
s?family=Roboto:regular,bold&subset=Latin">
<link rel="stylesheet" type="text/css" href="http://fonts.googleapis.com/
css?family=Fira Sans:regular,bold&subset=Latin">
<link rel="stylesheet" type="text/css" href="http://fonts.googleapis.com/
css?family=Roboto Mono:regular,bold&subset=Latin">
<!-- Icons Stylesheet -->
<link href="https://cdnjs.cloudflare.com/ajax/libs/font-awesome/6.4.2/css/
all.min.css" rel="stylesheet">
<!-- Customized Bootstrap Stylesheet -->
<!-- This file is downloaded from "https://getbootstrap.com/" online and
placed in the "css" folder. -->
<link href="css/bootstrap.min.css" rel="stylesheet">
```

Adding the Graph and Dashboard Components

Placed within the body tag, the components making the application are given in Listing 5-5, together with their description and purpose in the comments. This is also where the external scripts, i.e., chart.js and pop.js, are called. These are referenced at the end of the body to ensure that the document object model (DOM) is fully loaded before execution.

Listing 5-5. Adding the graph and dashboard components

```
<!-- Title to be displayed at the top of popup -->
<p id="title" class="mb-0">Network Traffic Monitoring and Analysis</p>

<!-- Chart Area -->
<div class="container-fluid pt-2">
   <div class="row g-2">
      <div class="col-sm-12 col-xl-3">
         <!-- Container for the chart -->
         <div class="bg-secondary rounded d-flex align-items-center
         justify-content-between p-3">
            <!-- Canvas for the chart -->
            <canvas id="myChart" height="250"></canvas>
         </div>
```

```
        </div>
     </div>
</div>

<!-- Dashboard Components - Predicted Metrics Cards -->
<div class="dashboard-container">
   <!-- Title for the Predicted Metrics -->
   <p class="dashboard-title">Predicted Metrics</p>
   <!-- Row for the Predicted Metric Cards -->
   <div class="dashboard-row">
      <!-- Card for Download Speed -->
      <div class="dashboard-card">
         <!-- Icon for Download Speed -->
         <i class="icon fas fa-circle-chevron-down"></i>
         <!-- Download Speed Metric to be displayed -->
         <span class="metric" id="results_downloadSpeed" style="color:
         #FF6384;">Hold on...</span>
         <!-- Description for Download Speed -->
         <p class="description">Download Speed (kbps)</p>
      </div>
      <!-- Card for Upload Speed -->
      <div class="dashboard-card">
         <!-- Icon for Upload Speed -->
         <i class="icon fas fa-circle-chevron-up"></i>
         <!-- Upload Speed Metric to be displayed -->
         <span class="metric" id="results_uploadSpeed" style="color:
         #2CD3E1;">Hold on...</span>
         <!-- Description for Upload Speed -->
         <p class="description">Upload Speed (kbps)</p>
      </div>
      <!-- Card for Latency -->
      <div class="dashboard-card">
         <!-- Icon for Latency -->
         <i class="icon fas fa-hourglass"></i>
         <!-- Latency Metric to be displayed -->
```

```
        <span class="metric" id="results_latency" style="color:
        #45FFCA;">Hold on...</span>
        <!-- Description for Latency -->
        <p class="description">Latency (ms)</p>
      </div>
    </div>
</div>

<!-- Dashboard Components - Network Performance Cards -->
<div class="dashboard-container" style="margin-top: 5px">
    <!-- Title for Network Performance -->
    <p class="dashboard-title">Network Performance</p>
    <!-- Row for the Network Performance Cards -->
    <div class="dashboard-row">
        <!-- Card for Network Activity -->
        <div class="dashboard-card">
            <!-- Icon for Network Activity -->
            <i class="icon fas fa-circle-nodes"></i>
            <!-- Network Activity Metric to be displayed -->
            <span class="metric" id="results_networkActivity" style="color:
            #FF6384;">Hold on...</span>
            <!-- Description for Network Activity -->
            <p class="description">Network Activity</p>
        </div>
        <!-- Card for QoS Score -->
        <div class="dashboard-card">
            <!-- Icon for QoS Score -->
            <i class="icon fas fa-star"></i>
            <!-- QoS Score Metric to be displayed -->
                <span class="metric" id="results_qosScore" style="color:
                #2CD3E1;">Hold on...</span>
            <!-- Description for QoS Score -->
            <p class="description">QoS Score</p>
        </div>
        <!-- Card for Average Latency -->
        <div class="dashboard-card">
```

```
    <!-- Icon for Average Latency -->
    <i class="icon fas fa-gauge"></i>
    <!-- Average Latency Metric to be displayed -->
    <span class="metric" id="results_averageLatency" style="color:
    #45FFCA;">Hold on...</span>
    <!-- Description for Average Latency -->
    <p class="description">Average Latency (ms)</p>
  </div>
 </div>
</div>

<!-- Fields for user inputs -->
<div class="container-fluid pt-1">
  <!-- Row for user input fields -->
  <div class="row g-2">
    <!-- Column for user input fields -->
    <div class="col-sm-12 col-xl-3">
      <!-- Background and styling for user input fields -->
      <div class="bg-secondary rounded d-flex align-items-center p-2">
    <!-- "Window Size" input field -->
        <div class="form-floating mb-1" style="padding-right: 5px;">
          <!-- Text input for "Window Size" -->
          <input type="text" class="form-control"
          id="floatingWindowSize" placeholder="" value="60">
          <!-- Label for "Window Size" -->
          <label for="floatingWindowSize" id="floatingLabels">Window
          Size</label>
        </div>
        <!-- "Prediction Time" input field -->
        <div class="form-floating mb-1" style="padding-left: 5px;">
          <!-- Text input for "Prediction Time" -->
          <input type="text" class="form-control"
          id="floatingPredictionTime" placeholder="" value="120"
          style="width: 240px;">
          <!-- Label for "Prediction Time" -->
```

```
            <label for="floatingPredictionTime"
            id="floatingLabels">Predict traffic for (s)</label>
        </div>
        <!-- Button for starting the monitoring and analysis processes,
        i.e., triggering the server -->
        <button id="monitorButton">Monitor & Analyze</button>
      </div>
    </div>
  </div>
</div>

<!-- Include scripts for chart.js and pop.js at the end of the body to
ensure DOM is fully loaded before execution -->
<script src="chart.js"></script>
<script src="pop.js"></script>
```

Styling the Components through Internal CSS

To give an appealing look to the application, along with enhancing its user-friendliness, the style tag is populated with the codes in Listing 5-6.

Listing 5-6. Styling the components through internal CSS

```
/* Styles for the body */
body {
background-color: #131F33;
color: #92ABCF;
font-family: "Roboto";
font-size: 15px;
width: 710px;
}
/* Styles for the body's background */
.bg-secondary {
background-color: #0F1724 !important;
}
```

```css
/* Styles for any primary text */
.text-primary {
color: #92ABCF !important;
}
/* s for the "Monitor & Analyze" Button */
#monitorButton {
display: flex;
width: 200px;
height: 30px;
align-content: right;
margin-top: 10px;
margin-bottom: 10px;
margin-left: 12px;
text-transform: uppercase;
flex-direction: row;
-moz-box-align: center;
align-items: center;
gap: 0.375rem;
border: medium none;
box-shadow: rgba(0, 0, 0, 0.2) 0px 3px 1px -2px, rgba(0, 0, 0, 0.14) 0px
2px 2px 0px, rgba(0, 0, 0, 0.12) 0px 1px 5px 0px;
font-weight: bold;
font-family: Fira Sans, Arial, Helvetica, sans-serif;
cursor: pointer;
transition: background-color 250ms cubic-bezier(0.4, 0, 0.2, 1) 0ms,
box-shadow 250ms cubic-bezier(0.4, 0, 0.2, 1) 0ms, border 250ms cubic-
bezier(0.4, 0, 0.2, 1) 0ms;
-moz-box-pack: center;
justify-content: center;
border-radius: 12px;
padding: 0.625rem 0.75rem;
font-size: 0.875rem;
color: rgb(0, 0, 0);
background-color: rgb(17, 236, 229);
}
```

```css
/* Styles for appearance of button upon hover */
#monitorButton:hover {
background-color: #bd1515;
color: white;
transition: .5s;
}
/* Styles for "Window Size" field */
#floatingWindowSize {
background-color: rgb(17, 236, 229);
color: rgb(0, 0, 0);
height: 50px;
}
/* Styles for appearance of the "Window Size" field when clicked */
#floatingWindowSize:focus {
border: 3px solid rgb(17, 236, 229) !important;
box-shadow: 0 0 3px rgb(17, 236, 229) !important;
-moz-box-shadow: 0 0 3px rgb(17, 236, 229) !important;
-webkit-box-shadow: 0 0 3px rgb(17, 236, 229) !important;
}
/* Styles for "Predict traffic for (s)" field */
#floatingPredictionTime {
background-color: rgb(17, 236, 229);
color: rgb(0, 0, 0);
height: 50px;
}
/* Styles for appearance of the "Predict traffic for (s)" field when
clicked */
#floatingPredictionTime:focus {
border: 3px solid rgb(17, 236, 229) !important;
box-shadow: 0 0 3px rgb(17, 236, 229) !important;
-moz-box-shadow: 0 0 3px rgb(17, 236, 229) !important;
-webkit-box-shadow: 0 0 3px rgb(17, 236, 229) !important;
}
```

```css
/* Styles for "Window Size" and "Predict traffic for (s)" fields */
#floatingLabels {
color: rgb(0, 0, 0);
font-weight: bold;
}
/* Styles for any input fields */
input:focus, textarea:focus, select:focus {
outline-offset: 0px !important;
outline: none !important;
}
/* Styles for the title of the popup */
#title {
padding-top: 2px;
padding-left: 15px;
text-align: left;
font-family: Roboto Mono;
font-weight: bold;
font-size: 14px;
color: #DFCCFB;
}
/* Styles for the title of each dashboard row */
.dashboard-title {
color: #DFCCFB;
margin-left: 10px;
margin-bottom: -5px;
font-size: 13px;
font-family: 'Roboto Mono';
}
/* Styles for a dashboard row */
.dashboard-row {
display: flex;
justify-content: space-between;
}
```

```css
/* Styles for a dashboard card */
.dashboard-card {
padding: 7px;
margin: 6px;
border-radius: 10px;
background-color: #101424;
position: relative;
width: 225px;
}
/* Styles for a dashboard container */
.dashboard-container {
background-color: #1d243f;
margin-top: 2px;
border-radius: 10px;
position: relative;
width: 100%;
}
/* Styles for a dashboard icon */
.icon {
position: absolute;
top: 10px;
right: 10px;
font-size: 25px;
color: #DFCCFB;
}
/* Styles for a dashboard metric */
.metric {
font-size: 18px;
}
/* Styles for a dashboard description */
.description {
font-size: 11px;
font-family: Fira Sans;
}
```

Adding the User Interface to Google Chrome

Now that the frontend is completely built, it is added to Google Chrome as shown in the following steps:

1. On the main menu of Google Chrome, click the three dots icon located on the top far right of the window, after which "Extensions" should be selected (Figure 5-5). This is followed by clicking on "Manage Extensions," which opens up Google Chrome Extensions.

Figure 5-5. *Accessing the extensions menu*

2. On the top left of the menu, the "Load Unpacked" option is selected, as shown in Figure 5-6.

Figure 5-6. *Loading an unpacked extension*

3. The project's root directory is then selected, as shown in Figure 5-7.

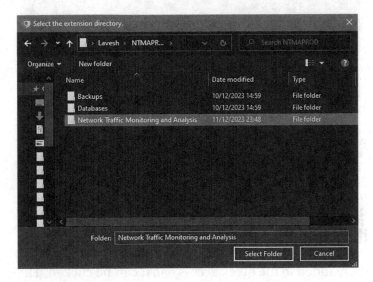

Figure 5-7. *Selecting the project's root directory*

4. This adds the application to the Extensions menu (Figure 5-8).

Figure 5-8. *NTMA application added to the Extensions menu*

5. The application can be pinned at the top right of Google Chrome for quick access by clicking on the puzzle icon, followed by the pin icon, as shown in Figure 5-9.

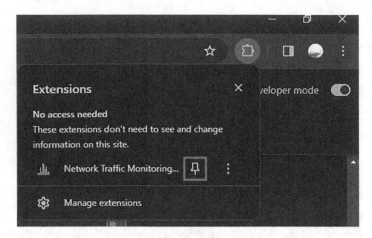

Figure 5-9. *Pinning the application for quick access*

Visualizing the User Interface

With the application added to the quick access bar, it can be clicked. The completed user interface is given in Figure 5-10.

Figure 5-10. *NTMA application completed user interface*

5.2.4 Building the Client Script

The client's logic is built with JavaScript and consists of a single file, pop.js, which primarily handles communication with the server through WebSockets [11], updates the UI with the information received, and orchestrates the real-time monitoring graph using Chart.js.

File Functionality

In this section, a breakdown of the pop.js file is given, whose functions are as follows:

1. Declare global variables for use throughout the script.

2. Query the Chrome tab using Chrome extension API:

   a. Retrieve information about the active and last focused-on tab.

   b. Retrieve the URL of the active tab.

3. Create a WebSocket connection to the server:

   a. Implement method for receiving messages from the server.

   b. Handle the connection using the open event.

   c. Parse and log messages.

4. Receive and handle messages from the server:

   a. Parse JSON messages containing real-time metrics and graph updates.

   b. Update HTML elements with the predicted metrics and network performance metrics.

   c. Update the Chart.js chart dynamically with the latest metrics.

5. Create an event listener for "Monitor & Analyze" button:

   a. Add an event listener to the button.

   b. Send a message to the server to start the monitoring process.

   c. Encapsulate "Window Size" and "Prediction Time" in a JSONObject sent to the server.

6. Create an event listener for the DOM:

    a. Add an event listener to the DOM.

    b. Create configurations for the three datasets and initialize the Chart.js chart object.

7. Implement a method to update the chart every second based on a specified label.

Creating the Script Structure

With the libraries already called and referenced in the popup.html file, the pop.js file can be populated by first creating the script structure, which consists of the following fundamentals:

- Declare the global variables.

- Query the active and last focused-on tab to fetch the current URL.

- Add an event listener on the DOM to be executed as soon as the page loads.

- Create a method to update the chart dynamically.

The skeleton for these functionalities is shown in Listing 5-7, together with the placement for further codes.

Listing 5-7. Creating the script structure

```
// Declare the global objects that are called throughout the script.
var monitorButton; // Variable for the "Monitor & Analyze" object.
var myChart; // Variable for the Chart.js object.

// Query the active and last focused-on tab to retrieve the current URL.
chrome.tabs.query({
    active: true,
    lastFocusedWindow: true,
},
    function(tabs) {
```

```
        <!-- Listing 5-9 is included here -->
    }
);

// Adding an event listener on the DOM to execute as soon as the
page loads.
document.addEventListener('DOMContentLoaded', function() {

        <!-- Listing 5-8 is included here -->

}, false);

// Method to update the chart every second.
function addData(chart, label, data, updateLabel) {

        <!-- Listing 5-10 is included here -->

}
```

Initializing and Configuring the Chart on Page Load

The DOMContentLoaded event is responsible for executing statements as soon as the HTML document's DOM is fully loaded. With the chart's area defined through the myChart tag, as specified in popup.html, it needs to be initialized with the proper configurations to accommodate three different datasets, with different settings. The following functions are then performed by adding an event listener on the whole document, which acts immediately on page load:

1. Get the DOM element for the "Monitor & Analyze" button, and assign it to the global variable monitorButton.

2. Create dataset configurations:

    i. Create three separate datasets for "Download Speed," "Upload Speed," and "Latency."

    ii. For each dataset, specify the label, background color, border color, border width, and point radius.

3. Create a data object that encapsulates all the dataset settings, thereby fed directly as a parameter to Chart.js.

4. Create an `options` object that styles the appearance of the Chart.js chart, with settings including responsiveness, aspect ratio, line tension, axis configurations, tick display settings, and legend styles.

5. Create a `config` object for the Chart.js chart where the type of the chart, the data to be included, and the specified options are all encapsulated.

6. Create a new Chart.js instance using the `myChart` identifier, thus binding all the settings to the HTML element. This indicates the object where the line chart is rendered.

Listing 5-8 gives the codes for the mentioned functionalities.

Listing 5-8. Initializing and configuring the chart on page load

```
// Get the DOM element for the "Monitor & Analyze" button.
monitorButton = document.getElementById('monitorButton');

// Create the dataset configuration for the "Download Speed" line chart.
const dataset_DS = {
    label: 'Download Speed (kbps)',
    backgroundColor: 'rgb(255, 99, 132)',
    borderColor: 'rgb(255, 99, 132)',
    borderWidth: 1.2,
    pointRadius: 1.6
};

// Create the dataset configuration for the "Upload Speed" line chart.
const dataset_US = {
    label: 'Upload Speed (kbps)',
    backgroundColor: 'rgb(44, 211, 225)',
    borderColor: 'rgb(44, 211, 225)',
    borderWidth: 1.2,
    pointRadius: 1.6
};
```

```javascript
// Create the dataset configuration for the "Latency" line chart.
const dataset_Lat = {
    label: 'Latency (ms)',
    backgroundColor: 'rgb(69, 255, 202)',
    borderColor: 'rgb(69, 255, 202)',
    borderWidth: 1.2,
    pointRadius: 1.6
};

// Create the data object that encapsulates all datasets into one.
const data = {
    datasets: [dataset_DS, dataset_US, dataset_Lat]
};

// Create the options object that is responsible for styling the graph's
appearance.
const options = {
    // Set the width to be responsive to the area allocated.
    responsive: true,
    // Disallow the aspect ratio to change.
    maintainAspectRatio: false,
    elements: {
        line: {
            tension: 0.35, // Add a slight curvature to the line graphs.
        },
    },
    // Modify x-axis configurations.
    scales: {
        x: {
            display: true, // Show x-axis.
            title: {
                display: true, // Show title of x-axis.
                text: 'Time (s)', // Change label of x-axis.
                color: '#DFCCFB', // Change color of x-axis label.
```

```
                    font: {
                        family: 'Fira Sans', // Change font of x-axis label.
                        size: 14, // Change font size of x-axis label.
                    },
                },
                ticks: {
                    display: true, // Show ticks on x-axis.
                    color: '#6b7e9b', // Change label color of x-axis tick.
                    font: {
                        family: 'Ubuntu', // Change font of x-axis tick.
                        size: 10, // Change font size of x-axis tick.
                    },
                    stepSize: 10, // Change the step size for ticks on x-axis.
                },
            },
            y: {
                display: true, // Show y-axis.
                title: {
                    display: true, // Show title of y-axis.
                    text: 'Value', // Change label of y-axis.
                    color: '#DFCCFB', // Change color of y-axis label.
                    font: {
                        family: 'Fira Sans', // Change font of y-axis label.
                        size: 14, // Change font size of y-axis label.
                    },
                },
                ticks: {
                    display: false, // Disable ticks on y-axis.
                    color: '#6b7e9b', // Change label color of y-axis tick.
                    font: {
                        family: 'Ubuntu', // Change font of y-axis tick.
                        size: 13, // Change font size of y-axis tick.
                    },
                },
            },
        },
    },
```

```
    plugins: {
        legend: {
            labels: {
                color: '#DFCCFB', // Change legend label color.
                font: {
                    family: 'Roboto Mono', // Change font of legend labels.
                }
            }
        }
    }
};

// Configuration object for the line chart.
const config = {
    type: 'line', // Set the chart type.
    data: data, // Set the data for the chart.
    options: options // Set the options for styling and customization.
};

// Create a new Chart instance using the configuration and attach it to the
canvas element with the id "myChart".
myChart = new Chart(document.getElementById('myChart'), config);
```

Real-Time WebSocket Communication and Dynamic Graph Updates

The primary functions of the client are to first connect to a server on a specified port before opening up the connection to allow a flow of messages. It can also send messages when necessary, but since the server is responsible for most of the computation in this application, the client mostly handles the receipt of messages from the server. A breakdown of the functions involved in this part is as follows:

1. Get the URL of the active tab and store it in a variable.

2. Initialize a time variable, which acts as a time counter for the graph.

3. Create constants to dynamically differentiate between the three graphs.

4. Establish a WebSocket connection on "ws://localhost:8080/", i.e., through port 8080, for real-time communication.

   i. To send a message, the `ws.open` event is used. In this case, the URL of the active tab stored earlier is sent to the server for testing the communication pathway.

   ii. To receive a message, the `ws.onmessage` event is used. This method is automatically triggered upon receipt of any message from the server. Since the server is programmed to send messages every second, this method is also executed every second. Once received, the message is filtered accordingly for the real-time values, predicted metrics, and network performance metrics, where these values are used to update the HTML elements and the graph.

5. Check for the receipt of a message related to any error occurred at the server side and generate an alert if so.

6. Trigger real-time graph updates using the `addData()` method, which takes as input the chart object, the timestep, the real-time value, and a tag for dictating which dataset is being populated among the three.

7. Create an event listener for the "Monitor & Analyze" button:

   i. Send a "Start Monitoring" message to the server using the `send` keyword on the `ws` object. The server is programmed to start its processes once this particular message is received.

   ii. Encapsulate "Window Size" and "Prediction Time" variables, which are fetched from their respective HTML elements, into a `JSONObject` variable, which is made using the `stringify()` method.

In short, Listing 5-9 orchestrates the initialization of objects, creates a WebSocket connection, handles incoming messages from the server, calls a method to update the graph in real time, and conveys a keyword to the server to initiate the monitoring process.

Listing 5-9. Real-time WebSocket communication and dynamic graph updates

```javascript
// Store the URL of the active tab in a variable.
var tempurl = tabs[0].url;

// Variable for iterating time on the graph.
var time = 0;

// Constants for updating graph data.
const updateDownload = "downloadTraffic";
const updateUpload = "uploadTraffic";
const updateLatency = "latency";

// Create a WebSocket connection to the server.
const ws = new WebSocket('ws://localhost:8080/');

// Handle the WebSocket connection "open" event.
ws.onopen = function() {
    // Log a message for when the extension has successfully connected to
    // the server.
    console.log('WebSocket Client Connected');

    // Send the URL of the active tab to the server as a test to check if
    // communication is possible.
    ws.send(tempurl);
};

// Handle received messages from the server.
ws.onmessage = function(e) {
    // Log any received message.
    console.log("Received from server: " + e.data);

    // Handle messages related to predicted metrics.
    // The following statement checks whether the message received contains
    // the predicted values.
```

```javascript
// These object names are set at the server side.
if (e.data.includes("predictedDownloadSpeed") || e.data.includes("predictedUploadSpeed") || e.data.includes("predictedLatency")) {
    // Parse the JSON message for predicted metrics.
    const msg = JSON.parse(e.data);

    // Extract the predicted values from the message using the same
    notation set at the server side.
    // Store these values in their corresponding variables.
    var predRX = msg.predictedDownloadSpeed;
    // Variable for the predicted Download Speed metric.
    var predTX = msg.predictedUploadSpeed;
    // Variable for the predicted Upload Speed metric.
    var predLat = msg.predictedLatency;
    // Variable for the predicted Latency metric.
    var scoreQoS = msg.predictedQoS;
    // Variable for the QoS Score.
    var deviceActivity = msg.deviceActivity; // Variable for the
    classification results, i.e., the device activity.
    var averageLatency = msg.averageLatency;
    // Variable for the average latency metric.

    // Update the HTML elements with the received metrics from
    the server.
    document.getElementById("results_downloadSpeed").
    innerHTML = predRX;
    document.getElementById("results_uploadSpeed").
    innerHTML = predTX;
    document.getElementById("results_latency").
    innerHTML = predLat;
    document.getElementById("results_qosScore").
    innerHTML = scoreQoS;
    document.getElementById("results_networkActivity").innerHTML =
    deviceActivity;
```

```
    // Round the average latency to two decimal places using the
    toFixed() method.
    document.getElementById("results_averageLatency").innerHTML =
    averageLatency.toFixed(2);
}

// Handle messages related to any error sent by the server.
// The following statement checks whether the message received contains
the "error" keyword.
// This keyword is set at the server side.
if (e.data.includes("error")) {
    // Generate an alert to let the user know that latency values
    cannot be measured since there is no connection.
    alert("Unable to measure latency due to no connection.");
}

// Parse the JSON message for real-time metrics and graph updates.
// This occurs every second, since the server sends a message to the
extension every second.
const msg = JSON.parse(e.data);
var RX = msg.downloadSpeed; // Store the real-time Download Speed
                              metric in a variable.
var TX = msg.uploadSpeed; // Store the real-time Upload Speed metric in
                            a variable.
var latency = msg.latency; // Store the real-time Latency metric in a
                             variable.

// Add data to the chart using the addData() method by parsing the
following parameters:
// 1. Chart object.
// 2. Timestep (relative to when the user clicks on the button - this
denotes t = 0).
// 3. Real-time value.
// 4. Tag for guiding the variable to the right set of values on
the chart.
```

```
    addData(myChart, time, RX, updateDownload); // Update the graph with
    the latest Download Speed variable.
    addData(myChart, time, TX, updateUpload);
    // Update the graph with the latest Upload Speed variable.
    addData(myChart, time, latency, updateLatency);
    // Update the graph with the latest Latency variable.

    // Increment time for the next data point.
    time = time + 1;
};

// Add an event listener to the "Monitor & Analyze" button.
monitorButton.addEventListener('click', function() {
    // Send a message to the server to start the monitoring process.
    // This is done so that when the server obtains this message, it knows
    the client is ready.
    ws.send("Start Monitoring");

    // Encapsulate "Window Size" and "Prediction Time" in a JSONObject.
    // These variables are obtained from the UI and are sent to the server
    since they are needed for analytics.
    const propertiesMessage = {
        windowSize: document.getElementById('floatingWindow
        Size').value,
        predictionTime: document.getElementById('floatingPredictionTi
        me').value
    };

    // Send the JSONObject to the server using the stringify() method.
    ws.send(JSON.stringify(propertiesMessage));
}, false);
```

Real-Time Chart Update Mechanism

This section refers to the addData() method, designed to update the Chart.js line chart graph at regular intervals of one second. The parameters taken as inputs are:

- The Chart.js instance, i.e., myChart. This remains the same for all three datasets.

- The label, representing the time instant at which the point is to be added.

- The real-time metric obtained from the server.

- The tag, which specifies which dataset to update among the three.

Moreover, with the y-axis ticks previously disabled through the options object, it is enabled once the first set of values is updated on the graph, to allow its seamless appearance. The codes for the addData() method are given in Listing 5-10.

Listing 5-10. Real-time chart update mechanism

```
// Determine for which dataset the method has been called through the tag
previously made.
// If the tag represents Download Speed.
if (updateLabel === "downloadTraffic") {
    // Push the label, i.e., the time instant.
    chart.data.labels.push(label);
    // Push the data, i.e., the real-time metric.
    chart.data.datasets[0].data.push(data);
}
// If the tag represents Upload Speed.
else if (updateLabel === "uploadTraffic") {
    // Push the data, i.e., the real-time metric.
    chart.data.datasets[1].data.push(data);
}
// If the tag represents Latency.
else if (updateLabel === "latency") {
    // Push the data, i.e., the real-time metric.
    chart.data.datasets[2].data.push(data);
}
```

213

```
// Update the chart options to display the y-axis ticks when this method is
first called.
// This is done since the y-axis ticks are disabled at first.
myChart.options.scales.y.ticks.display = true;

// Update the chart.
chart.update();
```

5.3 Server Program Structure for NTMA

In this section, the Node.js server, which is responsible for the following main actions, is elaborated:

1. **Import modules**: The file starts by adding modules using the `require` keyword. The libraries added help create the HTTP server, the WebSocket connection, read system information, read local databases, and implement the ML algorithms.

2. **Initialize global variables**: Declare and initialize the objects to be used throughout the script.

3. **Read locally stored databases**: The server is responsible for reading previously stored values from a local database. These values are stored in arrays for further use.

4. **Create an HTTP server**: An HTTP server is hosted on port 8080 and listens for incoming requests from a client requesting connection. This server hosts the WebSocket connection.

5. **Handle WebSocket connections**: The WebSocket server is created and bound to the HTTP server. It then listens for connections from a connecting client, obtains the parameters sent, and triggers the monitoring and analysis processes.

6. **Network monitoring**: The `systeminformation` library helps collect network data such as download speed, upload speed, and latency values at regular intervals of one second. Other metrics

are also calculated. These values are sent to the client through the WebSocket connection in real time. Once network traffic has been recorded for one minute, the monitoring process is stopped, and the prediction and classification algorithms are triggered, together with the method to calculate the QoS score.

7. **MLP prediction method**: The predictMLP method is used to make time-series predictions using an MLP algorithm, which involves training the model on both real-time values, which are added to another array of time-series data previously collected. This process also calculates the percentage accuracy of the algorithm and uses sliding windows for forecasting.

8. **QoS score calculation method**: The calculateQoS method is used to calculate the QoS score based on jitter and latency values. This score ranges from "Optimal" to "Moderate" to "Poor."

9. **KNN classification method**: The calculateKNN method is used to classify the device activity using the KNN algorithm. It reads a classification database, shuffles data, trains the KNN model, and predicts the current device activity based on the last data point measured through network monitoring.

On the same wavelength, the script also logs statements to provide information about the server's current operations, messages from the client, and the metrics measured in real-time, as well as the results obtained from the ML algorithms. An overview of the program structure for the server side is given in Figure 5-11.

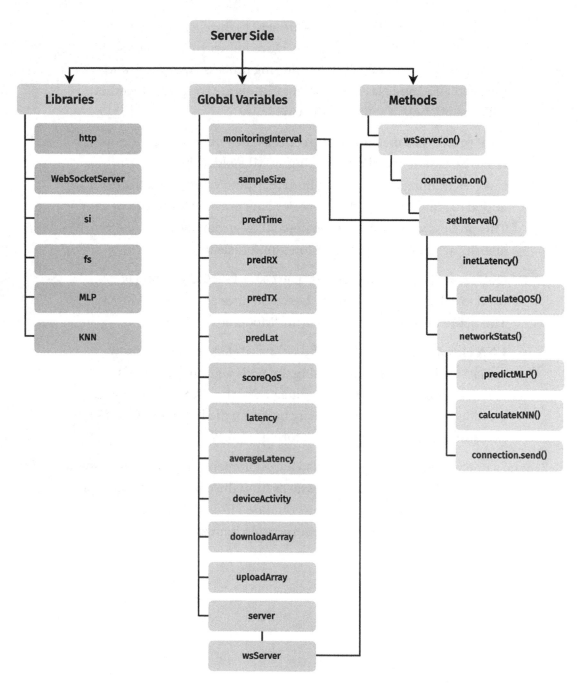

Figure 5-11. *Server-side program structure*

5.3.1 Libraries and Required Resources

Before coding the server's script, the libraries used are first downloaded locally through Node.js's console using the npm install keyword. The instructions for installing each particular module are given in the download links in Table 5-3.

Table 5-3. *Description of Libraries and Resources for the Server*

Library/Resource	Purpose	Download Link
http	Creates the HTTP server.	[12]
websocket	Binds onto the HTTP server for the WebSocket connection.	[1]
systeminformation	Provides access to network interfaces on the device to read network traffic.	[2]
fs	Works as a file system allowing access to files on the device.	[13]
mlp	Creates a perceptron to predict network traffic parameters.	[5]
ml-knn	Creates a KNN model to classify device activity.	[4]

5.3.2 Adding Libraries

To begin coding the engine of this application, the aforementioned libraries are added to the server's script using the require keyword one after the other at the top of the server.js file. This is shown in Listing 5-11.

Listing 5-11. Adding libraries

```
//Specify imports related to the server and monitoring process.
const http = require("http"); // Import the 'http' module for creating an
                              HTTP server.
const WebSocketServer = require("websocket").server; // Import the WebSocket
                                                    server module.
const si = require('systeminformation'); // Import the 'systeminformation'
library for obtaining system information about the device.
const fs = require('fs'); // Import the 'fs' module for working with the
                  file system.
```

```javascript
// Imports related to Machine Learning.
var MLP = require('mlp'); // Import the 'mlp' library, which is used for
                             Multilayer Perceptron prediction.
var KNN = require('ml-knn'); // Import the 'ml-knn' library which is used
                               for K-Nearest Neighbors classification.
```

5.3.3 Declaring Global Variables

Global variables, which are references throughout the script and used within some methods, are important to maintain state and share data across different parts of the script. They are thus declared and initialized right after importing the libraries, as shown in Listing 5-12.

Listing 5-12. Declaring global variables

```javascript
// Declare and initialize the global variables to be used across the
server's script.
var monitoringInterval; // Declare a variable for monitoring interval.
var sampleSize = 0; // Initialize a variable for sample size, use in the
                       data collection phase.
var predTime = 0; // Initialize a variable for prediction time at which the
                     user wants the prediction to be.
var predRX = 0; // Initialize a variable for prediction of download speed
                   in Kb/s.
var predTX = 0; // Initialize a variable for prediction of upload speed
                   in Kb/s.
var predLat = 0; // Initialize a variable for prediction of latency in ms.
var scoreQoS = ""; // Initialize a variable for the QoS score (Quality of
                      Service).
var latency = 0; // Initialize a variable for latency measurement in ms.
var averageLatency = 0; // Initialize a variable for calculating the
                           average latency in ms.
var deviceActivity = ""; // Initialize a variable for determining device
                            activity from "idle," "browsing," or "streaming."
```

5.3.4 Fetching Local Databases

To increase the efficiency of the algorithm, given that network traffic is collected for only one minute, previously collected values from the same network environment are loaded into two different arrays for download speed and upload speed. This means that these values will be collated with the traffic recorded when the "Monitor & Analyze" button is pressed. As a result, the MLP regression algorithm will be equipped with a significant dataset, enough to produce an accurate prediction. In this case, the downloadSpeed.txt and uploadSpeed.txt files are stored in the Databases folder, which in turn is located in the project's root directory.

The process is further explained as follows:

- The fs.readFileSync method is used to synchronously read the contents of the specified files.

- The files are .txt files that contain the elements separated by a line break.

- Regular expressions (regex) are used to split the file contents into values, which are fed into arrays.

- The resulting arrays, downloadArray and uploadArray, are used to initialize the historical network speed data for the ML models.

This is shown in Listing 5-13.

Listing 5-13. Fetching local databases

```
// Fetching some previously stored values from a locally stored database
for download speed.
const contents1 = fs.readFileSync("./Databases/downloadSpeed.txt",
'utf-8'); // Read the contents of the 'downloadSpeed.txt' file.
var downloadArray = contents1.split(/\r?\n/); // Split the contents into an
array, assuming each line represents a value.

// Fetching some previously stored values from a locally stored database
for upload speed.
const contents2 = fs.readFileSync("./Databases/uploadSpeed.txt", 'utf-8');
// Read the contents of the 'uploadSpeed.txt' file.
var uploadArray = contents2.split(/\r?\n/); // Split the contents into an
array, assuming each line represents a value.
```

5.3.5 Creating a WebSocket Server

A WebSocket server is commonly used in real-time applications, such as online gaming, financial platforms, and monitoring tools, where low latency is of utmost importance. WebSockets are increasingly being used in today's web applications due to their ability to allow seamless communication between the client and the server without constantly sending and receiving HTTP requests and answers. This is highly important in the case of data visualization dashboards, which require constant polling from a server, allowing quicker and more stable response times and improving performance and responsiveness. This translates into full-duplex communication channels over a single, long-lived connection. The bidirectional pathway is created through the following steps:

- Create an HTTP server using the `http.createServer` method while specifying the port number on which the server listens for incoming requests.

- Create an instance of the WebSocket server by referencing the variable used to import its module, then attach it to the HTTP server.

- Log messages concerning the server's status.

These processes are detailed in Listing 5-14.

Listing 5-14. Creating a WebSocket server

```
// Host a server connection on port 8080.
const server = http.createServer(); // Create an HTTP server instance.
server.listen(8080); // Listen on port 8080 for incoming HTTP requests.

console.log("Server started on port 8080..."); // Log a message indicating
that the server has started.
console.log("Waiting for connection..."); // Log a message indicating
                                          that the server is waiting for
                                          connections.

// Create a WebSocket server instance and attach it to the HTTP server
previously created.
const wsServer = new WebSocketServer({
    httpServer: server,
});
```

5.3.6 Listening for a Client Connection Request

This section involves several processes that pertain to the interactions between the client and the server as detailed in Figure 5-3. These are elaborated as follows:

- Listen for a connection request made by the client.

- Accept the connection.

- Obtain the "Window Size" and "Prediction Time" variables sent by the client through a JSON-formatted string.

- Read network data from the network interfaces of the device.

- Populate arrays for download speed, upload speed, and latency.

- Call the prediction and classification methods when one minute of network traffic is collected.

- Call the QoS calculation method when one minute of network traffic is collected.

- Send the results back to the client through WebSockets.

Listing 5-15 gives the structure of the processes involved in listening for and accepting a client request. The `connection.on` code block is left empty for now and is elaborated in the next section.

Listing 5-15. Listening for a client connection request

```
// Listen for a request for connection by a client through the
WebSocket server.
wsServer.on("request", function (request) {
// Accept a connection from a client.
    const connection = request.accept(null, request.origin);

    // Use the "message" keyword to accept incoming messages from
    the client.
    connection.on("message", function (message) {
        <!-- Listing 5-16 is included here -->
    });
```

```
    // Use the "close" keyword to handle client disconnection.
    connection.on("close", function (reasonCode, description) {
        // Log a message to indicate that the client has disconnected.
        console.log("Client has disconnected.");
         // Stop the monitoring process when the client disconnects.
        clearInterval(monitoringInterval);
    });
});
```

Getting Messages from the Client

The connection.on code block is triggered when any message is received by the server. The message is first logged on the console before being stored in a variable. Since the client is programmed to send the window size and prediction time variables' values when the "Monitor & Analyze" button is pressed, a check is performed through an if statement to detect the presence of the windowSize keyword in the message received. Once obtained, the monitoring process is started and a function is created to be triggered every second using the setInterval keyword. A time of 1000 ms is specified. This is shown in Listing 5-16, which is placed as indicated in Listing 5-15.

Listing 5-16. Getting messages from the client and creating a repeating function

```
// Get the messages obtained from the client.
// In this case, the first message received is the URL sent over by
the client.
// This ensures that the communication is up.
console.log("Message from Client:", message.utf8Data);

// Initialize a variable to contain messages sent over by the client.
var clientMessage = message.utf8Data;

// Check if the message sent by the client contains "windowSize" as one of
the variables in the JSONObject.
// This acts as the determinant to trigger the monitoring process.
// The receipt of the "windowSize" variable means that the client is
functioning properly.
```

```
if(clientMessage.includes("windowSize")) {
    const msg = JSON.parse(clientMessage); // Initialize a constant to
                                              parse the message.
    // Retrieve the "windowSize" and "predictionTime" values from the
    JSONObject.
    sampleSize = msg.windowSize; // "windowSize" value.
    predTime = msg.predictionTime; // "predictionTime" value.

    // Initialize a counter to know when enough network traffic has been
    collected.
    var counter = 0;

    // Initialize an array to contain latency items.
    var latencyArray = [];

    // Initialize a counter to know when enough latency values have been
    collected.
    var counterLatency = 0;

    // Create a function to be run every second using the
    setInterval method.
    // This method takes as input a function, and the time interval in ms.
    monitoringInterval = setInterval(function() {
        <!-- Listing 5-17 and Listing 5-18 are included here -->
    }, 1000) // Specify an interval of 1000 ms for the monitoring process.
}
```

Measuring Latency

The systeminformation library is accessed using the variable created upon adding the library. The following if statements are involved:

- A check is first performed to determine whether the latency value being returned by the library is null or not. The absence of a metric means that there is no internet connectivity. A message denoting the error is thus encapsulated in a JSON-formatted variable and sent to the client using the connection.send() method. This is used to generate an alert to the user on the frontend.

- If metrics are obtained, the respective array is filled.

- Once sixty values are collected, the `calculateQoS()` method is called to calculate the QoS score.

This is shown in Listing 5-17.

Listing 5-17. Measuring latency using the systeminformation library

```javascript
// Use the inetLatency() built-in method to read latency values.
si.inetLatency().then(data => {
    // Check if latency cannot be read, which would mean that there is no
    internet connection.
    if (data == null) {
        // Create an error message to be sent over to the client.
        const errorMessage = {
            error: "No Connection"
        };
        // Encapsulate the error message in a JSON-formatted String and
        send it over the WebSocket connection.
        connection.send(JSON.stringify(errorMessage));
        // Stop the monitoring process.
        clearInterval(monitoringInterval);
    // If connection is OK.
    } else {
        // Fill the latencyArray object with latency values using the
        counterLatency counter.
        latencyArray[counterLatency] = data;
        // Specify that the latency equals the data read from the Wi-Fi
        interfaces.
        latency = data;
    }
    // When a total of 60 values are recorded, one minute of traffic has
    been collected.
```

```
if (counterLatency == 59) {
    // Call the method to calculate the QoS based on the latency values
    recorded.
    scoreQoS = calculateQOS(latencyArray);
}
// Increment the counterLatency object.
counterLatency++;
})
```

Measuring Network Traffic and Sending Real-Time Metrics to the Client

Similar to the previous section, the systeminformation library is again used to read the number of packets received and sent per second, and stores this information in the RX and TX variables, respectively. Once sixty values are collected the following processes are triggered:

- Stop the monitoring process.

- Call the prediction method for making time-series prediction for download speed, upload speed, and latency, while catering for the return of NaN results. In the absence of a metric by the algorithms, the method is called again through a while loop. Once a metric is obtained, the loop is broken out of.

- Call the classification method to obtain the device activity by parsing download speed and upload speed as the input parameters.

- Create a variable to store the metrics and results obtained.

- Send this variable in a JSON-formatted message using the connection.send() method.

Now, if sixty values have not yet been collected, real-time metrics are also sent to the client using the connection.send() method. This is followed by incrementing a counter upon which the check for sixty values is performed. Listing 5-18 demonstrates these processes.

Listing 5-18. Measuring network traffic and sending real-time metrics to the client

```
// Fetch network data from Wi-Fi interfaces.
si.networkStats().then(data => {
    // Calculate the number of packets received per second, equivalent to
    the download speed.
    var RX = Math.round(((((data[0].rx_sec)/1000) + Number.EPSILON) *
    100) / 100;
    // Calculate the number of packets sent per second, equivalent to the
    upload speed.
    var TX = Math.round(((((data[0].tx_sec)/1000) + Number.EPSILON) *
    100) / 100;

    // Add newly read traffic to pre-loaded arrays using the push() method.
    // The pre-loaded arrays contain parameters stored in a time-series
    fashion.
    // This may increase the performance of the algorithm depending on the
    device activity.
    downloadArray.push(RX); // Add the download speed value to its pre-
                            loaded array.
    uploadArray.push(TX); // Add the upload speed value to its pre-
                          loaded array.

    // Log the values into console.
    console.log("Download Speed: " + RX + " KB/s");
    console.log("Upload Speed: " + TX + " KB/s");
    console.log("----------------------------");

    // When a total of 60 values has been recorded, the monitoring process
    is stopped and the ML algorithms are triggered.
    if (counter == 60) {
        // Stop the monitoring process using the clearInterval() method.
        clearInterval(monitoringInterval);
```

```
// Call the prediction method for making time-series prediction for
download speed.
// Ensure that the method is called again in case the algorithm
fails, and returns a NaN.
while (true) {
    predRX = predictMLP(downloadArray, parseInt
    (sampleSize), parseInt(predTime), 0.1, 0.01);
    if (!isNaN(predRX)) {
        // Break out of the loop if an actual number is obtained.
        break;
    }
}

// Call the prediction method for making time-series prediction for
upload speed.
// Ensure that the method is called again in case the algorithm
fails, and returns a NaN.
while (true) {
    predTX = predictMLP(uploadArray, parseInt
    (sampleSize), parseInt(predTime), 0.1, 0.01);
    if (!isNaN(predTX)) {
        // Break out of the loop if an actual number is obtained.
        break;
    }
}

// Call the prediction method for making time-series prediction for
latency.
// Ensure that the method is called again in case the algorithm
fails, and returns a NaN.
while (true) {
    predLat = predictMLP(latencyArray, parseInt
    (sampleSize), parseInt(predTime), 0.1, 0.01);
```

```
        if (!isNaN(predLat)) {
            // Break out of the loop if an actual number is obtained.
            break;
        }
    }

    // Classify the device activity by calling the
    // classification method.
    deviceActivity = calculateKNN(RX, TX);

    // Create a constant variable holding the predictions and
    // classifications results obtained.
    const predictionMessage = {
        downloadSpeed: RX,
        uploadSpeed: TX,
        predictedDownloadSpeed: predRX,
        predictedUploadSpeed: predTX,
        predictedLatency: predLat,
        predictedQoS: scoreQoS,
        deviceActivity: deviceActivity,
        averageLatency: averageLatency
    };
    // Encapsulate the metrics message in a JSON-formatted string and
    // send it over the WebSocket connection.
    connection.send(JSON.stringify(predictionMessage));
}
// If the number of values collected has not yet reached 60.
else {
    // Create a constant variable containing the network traffic values
    // to be sent to the client every second.
    // Add all 3 real-time values.
```

```
    const jsonMessage = {
        downloadSpeed: RX,
        uploadSpeed: TX,
        latency: latency
    };

    // Encapsulate the real-time message in a JSON-formatted String and
    // send it over the WebSocket connection.
    connection.send(JSON.stringify(jsonMessage));
    }
    // Increment the network traffic counter.
    counter = counter + 1;
})
```

5.3.7 Method for Time-Series Prediction with MLP Regression

As observed in Listing 5-18, the predictMLP() method is called for the three streams of data successively. The same logic applies to all and is depicted in Figure 5-12.

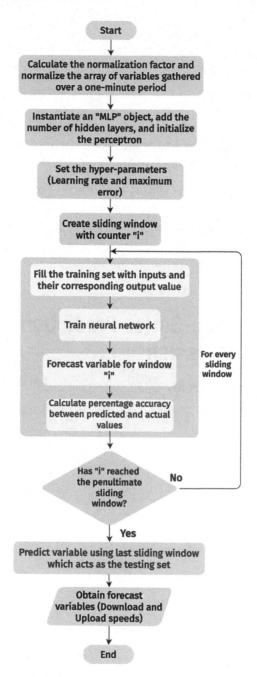

Figure 5-12. *Flowchart for time-series prediction using MLP*

Listing 5-19 gives the full code for the predictMLP() method.

Listing 5-19. Method for time-series prediction with MLP regression

```
// Create a method to make time-series predictions using multilayer
perceptron (MLP).
// Input parameters:
// - array: The time-series array to be predicted.
// - batchSize: The window size specified by the user on the UI.
// - pTime: The prediction time specified by the user on the UI.
// - learningRate: The learning rate for the MLP.
// - maximumError: The maximum error allowed during training.
function predictMLP(array, batchSize, pTime, learningRate, maximumError) {
    // Find the maximum and minimum values in the univariate stream.
    var min = Math.min.apply(Math, array); // Store the minimum value in a
                                            variable.
    var max = Math.max.apply(Math, array); // Store the maximum value in a
                                            variable.

    // Store the window size and prediction time values in new variables.
    var windowSize = batchSize;
    var pvalue = pTime;

    // Create the Perceptron using the "MLP" keyword and 1 input layer.
    var mlp = new MLP(windowSize, 1);
    // Add the number of hidden layers.
    mlp.addHiddenLayer((2 * windowSize) + 1);
    // Initialize the Perceptron.
    mlp.init();

    // Create the sliding window with a set of inputs and
    corresponding output.
    // Declare a variable for the inputs.
    var inputs = [];
    // Declare a variable for the outputs.
    var outputs = [];
```

```javascript
// Declare indexes to fill the "inputs" and "outputs" arrays.
var inputValue, outputValue;

// Initialize a counter for obtaining the actual values.
var counter = windowSize + pvalue - 1;

// Initialize an index for the windows. This ranges from 0 to
"windowSize-1".
var index = 0;

// Initialize a general index to move the window.
var valueFill = 0;

// Create sliding windows for window-sized-steps prediction.
for (let a = 0; a < (array.length - windowSize - pvalue + 1); a++) {
    // Fill the "inputs" array with the number of elements of the size
    of 1 window.
    for (let i = valueFill; i < (valueFill + windowSize); i++) {
        // Normalize each input such that it is found in the range
        of 0 to 1.
        inputValue = (array[i] - min) / (max - min);
        // Fill the "inputs" array with the normalized value depending
        on the index.
        inputs[index] = inputValue;
        // Increment the index.
        index = index + 1;
    }
    // Reset the index to 0 such that it can iterate from "0 to window"
    at the next sliding window.
    index = 0;
    // Move the general index to the next adjacent value.
    valueFill = valueFill + 1;
    // Normalize each output variable such that it is found in the
    range of 0 to 1.
    outputValue = (array[counter] - min) / (max - min);
    // Fill the "outputs" array with the normalized value. This array
    contains only 1 value.
```

```
    outputs[0] = outputValue;
    // Add the window-sized inputs and its corresponding output to the
    training set.
    mlp.addToTrainingSet(inputs, outputs);
    // Increment the main counter by 1.
    counter = counter + 1;
}

// Train the perceptron.
var learnRate = learningRate; // Specify the learning rate.
var error = Number.MAX_VALUE; // Initialize the error to a very
                                large value.
// Continue training until the error is less than the specified
maximum error.
while (error > maximumError) {
    // Train the MLP using the specified learning rate.
    error = mlp.train(learnRate);
}

// The algorithm has now been trained on real-time values.
// The sliding windows are again used to measure the % accuracy.
// A predicted value is generated for each window.

// Re-initialize previous variables such that they can be reused.
var inputs = [];
var inputValue;
var index = 0;
var valueFill = 0;

// Create a counter to measure the accuracy of the algorithm.
var accCounter = windowSize + pvalue - 1;
// Initialize a variable to calculate the average error.
var averageError = 0;

// Create sliding windows for window-sized-steps prediction.
for (let a = 0; a < (array.length - windowSize - pvalue + 1); a++) {
    // Fill the "inputs" array with the number of elements of the size
    of 1 window.
```

```javascript
for (let i = valueFill; i < (valueFill + windowSize); i++) {
    // Normalize each input such that it is found in the range
    of 0 to 1.
    inputValue = (array[i] - min) / (max - min);
    // Fill the "inputs" array with the normalized value depending
    on the index.
    inputs[index] = inputValue;
    // Increment the index.
    index = index + 1;
}
// Reset the index to 0 such that it can iterate from "0 to window"
at the next sliding window.
index = 0;
// Move the general index to the next adjacent value.
valueFill = valueFill + 1;
// Calculate the predicted value for this specific window using the
"classify" method.
var windowPrediction = mlp.classify(inputs);
// De-normalize the predicted value.
windowPrediction = windowPrediction * (max - min) + min;
// Initialize a variable to store the % error for each window.
var individualError = 0;
// Calculate each individual error by comparing with the
actual value.
if (array[accCounter] != 0) {
    individualError = ((Math.abs(windowPrediction -
    array[accCounter])) / (array[accCounter])) * 100;
} else {
    // In case the actual value is 0; to prevent division by 0
    individualError = ((Math.abs(windowPrediction -
    array[accCounter])) / (1)) * 100;
}
// Increment the accuracy counter by 1.
accCounter = accCounter + 1;
```

```
    // Add the individual error to a sum of errors.
    averageError = averageError + individualError;
}
```

```
// Calculate the average error by diving by the sum of errors by number
of predictions made.
averageError = averageError / (array.length - windowSize - 1);
// Log a message for the total average error.
console.log("Total Average Error: " + averageError + " %");
// Calculate the percentage accuracy.
var percentageAccuracy = 100 - averageError;
// Log the percentage accuracy.
console.log("Percentage Accuracy: " + percentageAccuracy  + " %");
```

```
// Create an array for the last "windowSize" values.
var pred = [];
// Create a starting index for the last sliding window in the dataset.
var arrayIndex = 0;
// Create an ending index for the last sliding window in the dataset.
var lastWindowIndex = array.length - windowSize;
// For the last possible sliding window.
for (let i = lastWindowIndex; i < array.length; i++) {
    // Fill the last sliding window array with the respective
    normalized values.
    pred[arrayIndex] = (array[i] - min) / (max - min);
    // Increment the counter by 1.
    arrayIndex = arrayIndex + 1;
}
```

```
// Calculate the predicted value for this specific window using the
"classify" method.
var mlpPrediction = mlp.classify(pred);
// De-normalize the predicted value.
mlpPrediction - (mlpPrediction * (max   min) ı min);
// Format the predicted value to 2 decimal places.
var predictedValue = Math.round((mlpPrediction + Number.EPSILON) *
100) / 100;
```

```
// Log the formatted predicted value.
console.log("Predicted value after " + pvalue + " seconds: " +
predictedValue);
console.log("----------------------------");
// Clear the training dataset.
mlp.clearTrainingSet();
// Reset the weights learned.
mlp.resetWeights();
// Return the final predicted value.
return predictedValue;
}
```

5.3.8 Method for Calculating the QoS Score

This section refers to the calculateQOS() method that is called in Listing 5-17. To measure the QoS score, jitter is first calculated based on the recorded latency samples, as per Equation 5-1, where n represents the number of samples, x is the sample value, and i is the order of collected values, starting from zero.

$$Jitter = \frac{\sum_{i=1}^{n}(x_i - x_{i-1})}{n-1} \tag{5-1}$$

Subsequently, the average latency observed within the sixty-second duration is computed. The QoS score is established using three thresholds, and calculated by comparing the average latency with jitter. A device is considered to be displaying "Poor" QoS if the jitter is greater than 50 percent of the average delay. "Moderate" QoS is assigned when the same result falls between 30 percent and 50 percent. Ultimately, the device is said to be operating at "Optimal" QoS if jitter is less than 30 percent of the average latency. These criteria are established to define various degrees of service quality, based on the average latency toward specific servers in the location where the application is being used. This varies from country to country and network to network. Such thresholds are often fine-tuned through iterative testing and validation based on factors including user feedback, industry standards, and service-level agreements. For example, a telecommunications operator might set the criteria for QoS based on the best possible latency obtained toward a particular country, through different networking routes. The QoS score determination process is summarized in Figure 5-13.

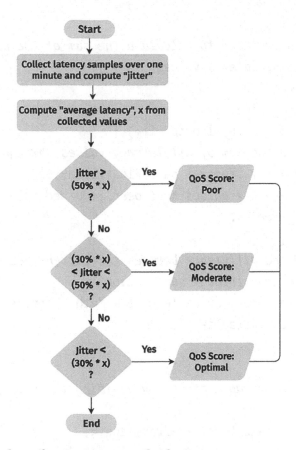

Figure 5-13. *Flowchart for QoS score calculation*

The code for this process is given in Listing 5-20.

Listing 5-20. Method for calculating the QoS score

```
// Create a method to calculate a QoS score depending on jitter and
latency values.
// Input parameters:
// - array: The time-series array to be parsed.
function calculateQOS(array) {
    // Initialize a variable to store the final score.
    // The score ranges from "Poor" to "Moderate" to "Optimal."
    var score = "";
    // Initialize a variable to calculate the total sum of all
    latency values.
```

```javascript
var sumLatency = 0;
// Initialize a variable to calculate the sum of the differences
between successive latency values.
var sumDifference = 0;
// Iterate through the array.
for(let i = 0; i < array.length; i++) {
    // Calculate the sum of all latency values through addition.
    sumLatency = sumLatency + parseFloat(array[i]);
    // When at least 2 values have been recorded, start calculating the
    sum of differences.
    if (i > 0) {
        // Add the difference between the current value and the
        previous one through subtraction.
        sumDifference = sumDifference + Math.abs(parseFloat(array[i]) -
        parseFloat(array[i-1]));
    }
}
// Calculate the average latency by dividing the sum by the number
of values.
averageLatency = sumLatency / array.length;
// Calculate jitter by dividing the sum of differences by the number
of values.
var jitter = sumDifference / array.length;

// Log the average latency.
console.log("Average Latency: " + averageLatency + " ms");
// log the jitter.
console.log("Jitter: " + jitter + " ms");

// If jitter is 50% of the average latency.
if (jitter > (0.5 * averageLatency)) {
    // Assign "Poor" to the QoS Score.
    score = score + "Poor";
    // Log the QoS Score.
    console.log(score);
}
```

```
// If jitter is between 30% to 50% of the average latency.
else if ((jitter > (0.3 * averageLatency)) && (jitter < (0.5 *
averageLatency))) {
    // Assign "Moderate" to the QoS Score.
    score = score + "Moderate";
    console.log(score);
// If jitter is less than 30% of the average latency.
} else {
    // Assign "Optimal" to the QoS Score.
    score = score + "Optimal";
    // Log the QoS Score.
    console.log(score);
}
// Return the final QoS Score.
return score;
}
```

5.3.9 Method for Classifying the Device Activity

With regard to identifying the type of activity exhibited by the device, a pre-classified dataset is required containing the download speed, upload speed, and the corresponding category of the activity, as elaborated in Chapter 4. A sample dataset is provided along with the book. However, to improve the reliability of the classification method for various forms of network activity, it is recommended to conduct supplementary testing on a larger dataset, which can be collected using the same application. By feeding the dataset a wide range of network traffic samples that reflect different types of activities, the algorithm can enhance its ability to generalize and accurately differentiate between distinct forms of network behavior. The procedure entails the acquisition and categorization of a diverse array of network traffic data to guarantee a thorough representation of potential scenarios experienced in real-world networks. The technical procedure involving data preparation and classification using the KNN algorithm is summarized in Figure 5-14.

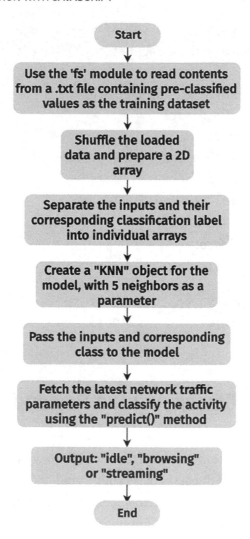

Figure 5-14. *Flowchart for classifying device activity*

The code for this process is given in Listing 5-21.

Listing 5-21. Method for classifying the device activity with KNN

```
// Create a method to calculate the device activity through classification.
// Input parameters:
// - dspeed: The download speed array.
// - uspeed: The upload speed array.
```

```
function calculateKNN(dspeed, uspeed) {
    // Create a temporary array to store data from the classification
    database.
    let temp = [];

    // Read the contents of the classification database file by specifying
    the location of the file.
    const contents0 = fs.readFileSync("./Databases/classification.txt",
    'utf-8');
    // Split the contents into an array based on line breaks.
    var classificationDatabase = contents0.split(/\r?\n/);

    // Loop through the classification database and populate the
    temporary array.
    for (let i = 0; i < classificationDatabase.length; i++) {
        // Add an array within the temporary array.
        temp.push([]);
        // Parse and store each entry as an array in the temporary array.
        temp[i] = [parseFloat(classificationDatabase[i].split(",")[0]),
        parseFloat(classificationDatabase[i].split(",")[1]),
        parseFloat(classificationDatabase[i].split(",")[2])];
    }

    // Shuffle the elements of the array.
    let data_array = temp;
    // The callback function passed to sort generates a random number
    between -0.5 and 0.5 for each pair of elements.
    // This results in a random order when sorting.
    data_array = data_array.sort(() => Math.random() - 0.5);
    // Initialize the shuffled array with the original array.
    let shuffledArray = data_array;

    // Initialize an empty array to store input data.
    let input = [];
    // Initialize an empty array to store output data.
    let output = [];
```

```javascript
// Loop through each element in the shuffled array.
for (let i = 0; i < shuffledArray.length; i++) {
    // Check if the current element does not include undefined values.
    if (!shuffledArray[i].includes(undefined)) {
        // Loop through each attribute (column) in the first row of the
        shuffled array.
        for (let j = 0; j < shuffledArray[0].length; j++) {
            // Check if the current attribute is at index 2.
            if (j === 2) {
                // Assign the value at index 2 to the output array.
                output[i] = shuffledArray[i][2];
            } else {
                // Create a new array within the input array if not
                already present.
                input.push([]);
                // Assign the value at the current attribute index to
                the input array.
                input[i][j] = shuffledArray[i][j];
            }
        }
    }
}

// Initialize a new array to store the values of the input array.
var newInputs = [];
// Loop through each element in the shuffled array.
for(let a = 0; a < shuffledArray.length; a++) {
    // Assign the values from the input array to the newInputs array.
    newInputs[a] = input[a];
}

// Create a new KNN (K-Nearest Neighbors) model with the specified
parameters:
// - newInputs: The input data for training the model.
// - output: The corresponding output classes for the training data.
// - { k: 5 }: An options object setting the number of neighbors (k) to
consider during classification.
```

242

```javascript
var knn = new KNN(newInputs, output, { k: 5 });

// Initialize a variable to count the number of classification errors.
var errorCount = 0;
// Iterate through each data point in the training set.
for (let i = 0; i < newInputs.length; i++) {
    // Predict the class using the KNN model for one particular the
    current input.
    let predictedValue = knn.predict(newInputs[i]);
    // Retrieve the actual class from the training set.
    let actualValue = output[i];
    // Check if the predicted class is different from the actual class.
    if(predictedValue !== actualValue) {
        // Increment the error count if there is a misclassification.
        errorCount++;
    }
}

// Calculate the KNN classification accuracy by finding the fraction of
errors over the number of inputs.
var knnAccuracy = 100 - ((errorCount / newInputs.length) * 100);
// Log the KNN classification accuracy.
console.log("KNN Classification Accuracy: " + knnAccuracy + " %");

// User input data to be classified (from browser in this context).
let feed = [dspeed, uspeed];

// Predict the class corresponding to the input data using the trained
KNN model.
let result = knn.predict([feed]);
// Initialize a variable for the classification result.
var classification = "";

// Log the input data and predicted class to the console.
console.log("Inputs: " + feed.toString() + "  Class: " + result.
toString());
```

```javascript
    // Use a switch statement to map the predicted class to a human-
    readable label.
    // If the result is 0, the label is set to "Idle".
    // If the result is 1, the label is set to "Browsing".
    // If the result is 2, the label is set to "Streaming".
    switch (parseInt(result)) {
        case 0:
            classification = classification + "Idle";
            break;
        case 1:
            classification = classification + "Browsing";
            break;
        case 2:
            classification = classification + "Streaming";
            break;
    }
    // Log the human-readable label classification result.
    console.log("Classification of value: " + classification);
    // Return the human-readable label classification result.
    return classification;
}
```

5.4 NTMA Application Testing and Deployment

With both the client and server built, the NTMA application is ready to be run. In this section, it is tested in a standard home environment with several scenarios. To simulate poor network conditions, a bandwidth allocation cap is deliberately placed on the device, in turn producing high latency and jitter. Figures 5-15 to 5-19 demonstrate the NTMA application in the described scenarios.

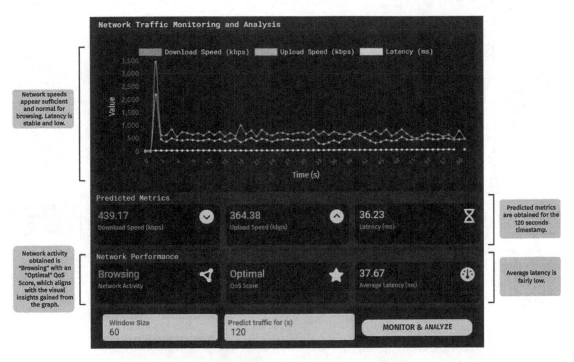

Figure 5-15. *Browsing on Facebook and Google News in an optimal environment*

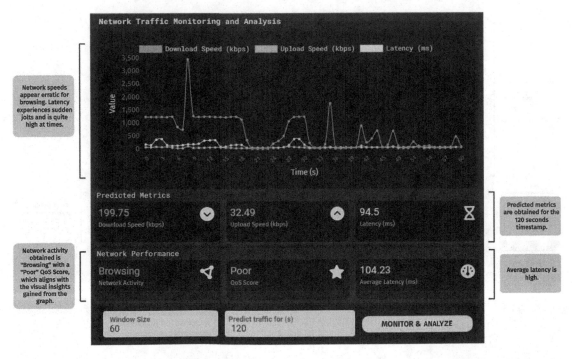

Figure 5-16. *Browsing on Facebook and Google News in a poor environment*

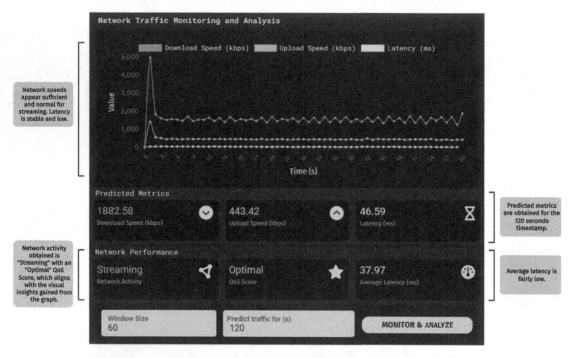

Figure 5-17. *Streaming on Twitch in an optimal environment*

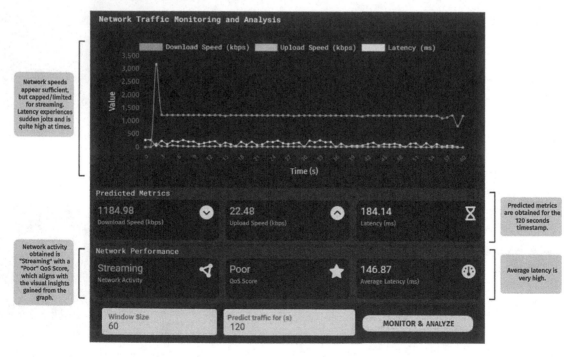

Figure 5-18. *Streaming on Twitch in a poor environment*

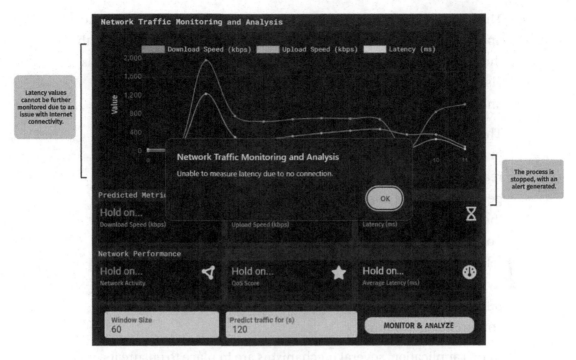

Figure 5-19. *Generation of an alert when connectivity is unavailable*

5.5 Summary

Chapter 5 was geared toward the application development of the NTMA framework with JavaScript based on concepts learned in previous chapters. The system model was first described, followed by the constituents and interconnections between client and server files, with specific emphasis on their individual characteristics and protocols. The user interface was built using a combination of HTML and CSS with user-friendliness as a foundation. The two machine learning models involved in this application were elaborated upon in code with the methods given for time-series prediction, classification of the device activity, and computation of the QoS score. The key takeaways are given here:

- The process of developing NTMA applications using JavaScript entails the creation and execution of client and server paradigms that are responsible for the monitoring and analysis of network traffic.

- The NTMA system model comprises client files, executed within Google Chrome as a browser extension, and server programmes executed on Node.js servers.

- The network monitoring process is triggered by the client side of the application through a button before seamlessly communicating with the server side for the reception of network traffic data. The server is also responsible for processing this data and making classification and prediction analysis through machine learning algorithms.

- The application is designed to adapt to ever-changing scenarios in terms of predictive analytics by combining data collected in real time with previously stored data. Since the computation is performed after a minute, which can also be adjusted accordingly, the results are obtained for the particular network scenario at that point in time. However, it is advised to use an appropriate dataset related to the network environment being analyzed. Moreover, the hyperparameters for MLP require fine-tuning to ensure optimal performance in changing ecosystems, in turn necessitating comprehensive testing.

- In this application, several mechanisms are in place to ensure its scalability in large-scale networks. First, it is designed to be modular and distributed, allowing for horizontal scalability by adding more Node.js servers as the network size increases.

- The use of port numbers allows for efficient data processing techniques such as parallelization and stream processing, thereby handling large volumes of network traffic in real time, where each client is handled independently.

- Cloud-based infrastructure can be leveraged to dynamically allocate resources and ensure optimal performance and efficiency.

In the next chapter, the video quality assessment (VQA) application is broken down with regards to its functionalities and codes.

5.6 References – Chapter 5

1. B. McKelvey, "websocket," npm, Apr. 14, 2021, `https://www.npmjs.com/package/websocket` (accessed Dec. 23, 2023).

2. S. Hildebrandt, "systeminformation," npm, Dec. 22, 2023, `https://www.npmjs.com/package/systeminformation` (accessed Dec. 23, 2023).

3. chart.js, "Chart.js | Open Source HTML5 Charts for Your Website," Chartjs.org, 2019, `https://www.chartjs.org/` (accessed Dec. 23, 2023).

4. ml.js, "ml-knn," npm, Jun. 29, 2019, `https://www.npmjs.com/package/ml-knn` (accessed Dec. 23, 2023).

5. U. Biallas, "mlp," npm, Jun. 30, 2016, `https://www.npmjs.com/package/mlp` (accessed Dec. 23, 2023).

6. Google, "Google Fonts," Google Fonts, 2019, `https://fonts.google.com/` (accessed Dec. 23, 2023).

7. Font Awesome 5, "Font Awesome 5," Fontawesome.com, 2017, `https://fontawesome.com/` (accessed Dec. 23, 2023).

8. M. Otto, "Bootstrap," Bootstrap, 2022, `https://getbootstrap.com/` (accessed Dec. 23, 2023).

9. "bootstrap," cdnjs, `https://cdnjs.com/libraries/bootstrap/5.0.0` (accessed Mar. 26, 2024).

10. "Chart.js," cdnjs, `https://cdnjs.com/libraries/Chart.js/3.8.0` (accessed Mar. 26, 2024).

11. mdn web docs, "The WebSocket API (WebSockets)," MDN Web Docs, Nov. 28, 2019, `https://developer.mozilla.org/en-US/docs/Web/API/WebSockets_API` (accessed Dec. 23, 2023).

12. Node.js, "HTTP | Node.js v18.8.0 Documentation," nodejs.org, `https://nodejs.org/api/http.html` (accessed Dec. 23, 2023).

13. Node.js, "File System | Node.js v18.1.0 Documentation," nodejs.org, `https://nodejs.org/api/fs.html` (accessed Dec. 23, 2023).

CHAPTER 6

Video Quality Assessment Application Development with JavaScript

This chapter continues the practical journey by presenting the application development aspect of the video quality assessment (VQA) model with JavaScript. To obtain a real-time mean opinion score (MOS) for video streaming quality, a no-reference (NR) image quality assessment (IQA) metric is employed. In an era where content is consumed in massive amounts in a matter of seconds, no original feed is usually available. With a live video as input, the algorithm presented in this book takes continuous screenshots, which are fed sequentially into the algorithm. Blocking, ringing, blurring, and noising are examples of artifacts against which the blind/reference-less image spatial quality evaluator (BRISQUE) is applied. This chapter therefore elaborates on how JavaScript is used to build a robust, reliable, and real-time VQA framework.

6.1 System Model for VQA

In this section, the system model for the VQA framework is elaborated. It consists of a three-layer architecture involving a client, backend server on Node.js, and local Java servlet hosted using Apache Tomcat. For testing purposes, the three systems are run on the same device. The end user is thus presented with a dashboard that has the following functionalities:

1. Monitor real-time video quality metrics directly in the web browser through a browser extension.

© Tulsi Pawan Fowdur, Lavesh Babooram 2024
T. P. Fowdur, L. Babooram, *Machine Learning For Network Traffic and Video Quality Analysis*,
https://doi.org/10.1007/979-8-8688-0354-3_6

2. Obtain a live video quality score ranging from 1 to 100 of the video stream being watched. Based on the terminology used by multiple vendors in the VQA sphere [1], the score is termed a live mean opinion score (MOS) where a lower value reflects a better objective quality. Likewise, a higher value points to higher levels of distortion.

3. Obtain a dominant quality rating summarizing the video quality performance over the frames analyzed.

These capabilities are coded in a way that video metrics are requested from the server, and hence the servlet, for a preset number of frames. The browser extension is powered by JavaScript (JS), with the backend being a Node.js server. In this case, the latter is simply responsible for being the intermediary between the client and the servlet. The servlet acts as the main VQA block, taking as input the image screenshot by the web browser, which passes through the Node.js server and is finally picked up by the servlet. HTML and CSS make up the frontend, while the extension and server send back-and-forth messages in JSON format. To trigger the servlet, an HTTP POST request is made by specifying the index of the screenshot taken. Upon request by the Node.js server, the servlet returns a JSON response containing the MOS score after a few seconds of processing. The server then relays this information to the client, which updates the GUI. The information received by the client also involves an average of the MOS score over the different analyzed frames, as well as a dominant quality rating, which signifies which rating has been obtained the most during a particular session. The ability to rapidly monitor video quality and produce several metrics in a matter of seconds is one of the prominent features of this VQA application, which is also packaged in a simple browser extension. Four files are required for the browser extension to take continuous screenshots and communicate with the Node.js server, as shown in the complete system model in Figure 6-1.

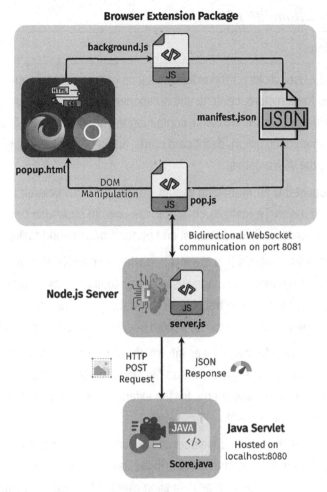

Figure 6-1. *Complete VQA system model*

6.1.1 Components and Functionalities

The WebSocket [2] library caters for the doubled communication framework between the Node.js server and the browser extension using messages encapsulated in JSON format. The functionalities of the four client files, Node.js server, and the servlet are summarized in Table 6-1.

Table 6-1. *File Functionalities for VQA*

File	Functions
popup.html	When the user clicks on the extension's icon in the web browser's toolbar, this file dictates the structure, content, and components in the dashboard that appears. It is thus referred to as the frontend, containing the user interface (UI) elements that make up the monitoring graph, dashboard cards, and a button for triggering the server and starting the VQA process.
background. js	It is accountable for managing events, states, and activities related to various parts of the extension. Typically, such scripts are used to supervise persistent interactions, manage communication patterns, and perform background tasks.
pop.js	It provides the routines that are invoked when UI components are interacted with and activated. Therefore, it manages user interactions and any dynamic information inside the popup, as well as the transmission of data to and from the server. Upon reception of output messages from the server, it updates the document object model (DOM).
manifest. json	As a file in JSON format, it gives information on scripts to be used and privileges that control tab usage. The manifest.json file also includes the name, version, permissions, actions, and icon image source for the extension. It thus gives the browser information concerning the extension.
server.js	Written in JS, the server.js file acts as the backend server, responsible for being the middleman between the client and the Java servlet. In this application, upon successful connection with an extension client, it first receives chunks of a screenshot taken by the client, which it compiles into a PNG image. Then, depending on the website being visited, it crops the image to only contain the frame of the video being watched using the Sharp library [3]. This frame is then stored locally, also as a PNG file. It then issues an HTTP POST request to the servlet and waits for a response containing the MOS score, which it relays back to the client. The server.js file contains several libraries that allow this seamless interaction to take place.
Servlet Package (VQA Block)	The servlet package is coded using Java in the Eclipse IDE. It is hosted on localhost using a downloaded Apache Tomcat server and programmed to receive HTTP POST requests. When triggered, the servlet retrieves the image previously stored by the Node.js server and puts it through a series of processing steps, i.e., the VQA block. The output is an MOS score, ranging from 1 to 100, which the servlet outputs to the server in a JSON format.

In short, when the user visits a video streaming website, such as YouTube or Twitch, the extension can be used to monitor the real-time video quality of the video being watched through an MOS score. This occurs through a series of screenshots taken as soon as the previous one is processed. The user simply clicks on the "Monitor" button and analyzes the dynamic graph, which is created using Chart.js [4].

6.1.2 Prediction of an MOS Score for Video Quality

In this framework, the screenshot taken by the client is passed through a series of processing steps before generating the MOS score. This score is a representation of the quality of the video being streamed. The application is suited for video players with adaptive bit rate (ABR) streaming. The aim here is to obtain both a score and a rating that can be understood by the average user. Every screenshot taken is labeled at the server side, since this same image is cropped and taken by the local servlet by specifying its file path. The parameter sent through the HTTP POST request dictates which screenshot is taken. The Node.js server is programmed to instantaneously request the client for another screenshot when a response has been obtained from the servlet. In doing so, it also sends the results. The Node.js server thus acts as the intermediary between the client and the servlet, thus performing operations such as cropping the image, and handling HTTP calls. This removes any processing load on the client, in turn preventing any lags and stuttering at the client side when viewing the video.

6.1.3 VQA Application Layout

In this application, the popup gives the user the ability to monitor real-time video quality metrics directly in the web browser. A live graph shows the metrics, which can be dynamically adjusted to only view the desired parameter. Once the extension is opened, it automatically searches for the Node.js server to connect to. The "Monitor" button acts as the trigger for the VQA process to start. The MOS score represents the quality score yielded by the servlet for one particular frame. When this button is clicked, a screenshot of the web browser is taken and sent to the Node.js server, where it is processed and fed to the servlet. Both a live and a dominant quality metric are available to the end user. The former represents the quality rating of each frame as being output by the servlet in real time, while the latter is an indication of the quality that has been obtained the most during a particular session. Figure 6-2 shows the completed browser extension.

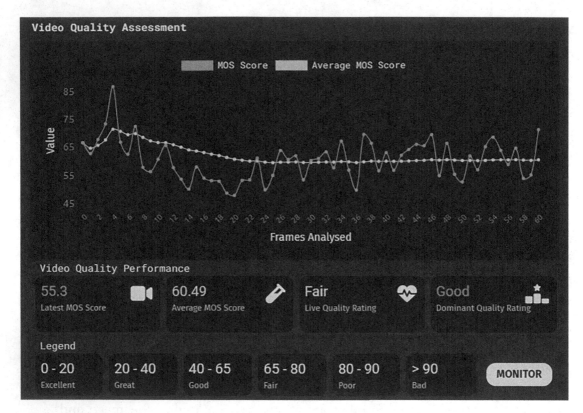

Figure 6-2. *VQA application layout*

6.1.4 Client–Server–Servlet Interaction

Figure 6-3 shows the stepwise three-way communication framework between the browser extension, the Node.js server, and the hosted Apache Tomcat Java servlet.

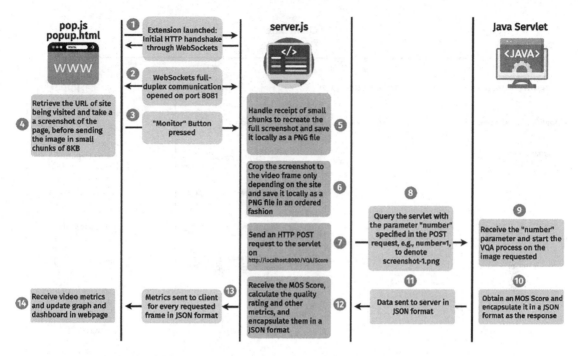

Figure 6-3. *Client–server–servlet interaction for VQA*

6.2 Client Program Structure for VQA

In this section, the client side, i.e., frontend, is described through the four fundamental files. Figure 6-4 gives an overview of the program structure for the client side.

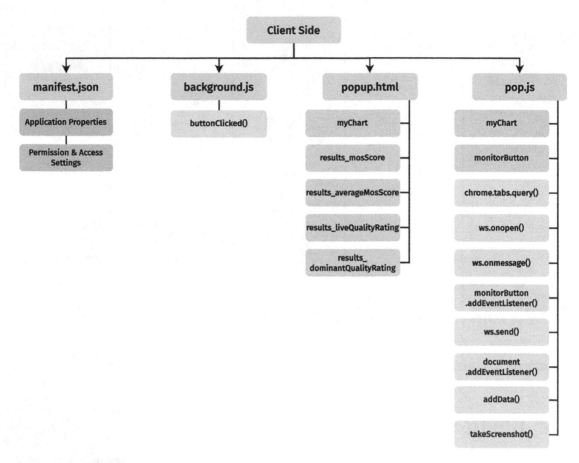

Figure 6-4. *Client-side program structure*

6.2.1 Configuring Extension Settings and Permissions

This section concerns the manifest.json file that contains metadata, allowing Google Chrome to add the browser extension to its list. The code for this file is given in Listing 6-1. The icons can be either added to the same file directory as the manifest.json file, or omitted in case the user prefers the default icon.

Listing 6-1. Configuring manifest.json with extension settings and permissions

```
{
  "manifest_version": 3,  // Manifest format version for Chrome extensions.

  "name": "Video Quality Assessment",  // Name of the Chrome extension.
  "version": "1.0",  // Version number of the Chrome extension.
```

```
  "icons": {
    "16": "icon16.png",    // Icon for 16x16 pixel size.
    "48": "icon48.png",    // Icon for 48x48 pixel size.
    "128": "icon128.png"   // Icon for 128x128 pixel size.
  },

  "background": {
    "service_worker": "background.js"  // Background script, implemented as
                                          a service worker.
  },

  "permissions": [ // List of permissions required by the extension.
    "activeTab",   // Allows extension to interact with the currently
                      active tab.
    "scripting",   // Permits the injection of scripts into web pages.
    "webNavigation",  // Enables the extension to receive events in the web
                         navigation lifecycle.
    "webRequest",  // Grants the extension access to observe and analyze
                      network requests.
    "storage",  // Provides access to the browser's storage APIs for data
                   persistence.
    "declarativeNetRequest",  // Allows the extension to declare a set of
                                 rules for web requests.
    "declarativeNetRequestFeedback",  // Permits the extension to provide
                                         feedback on web requests.
    "tabs"  // Enables the extension to query and manipulate browser tabs.
  ],

  "action": {
    "default_popup": "popup.html",  // Default popup HTML file when the
                                       extension icon is clicked.
    "default_title": "Video Quality Assessment"  // Default title for the
                                                     extension icon.
  }
}
```

6.2.2 Configuring the Background Script

The background script functions as an overlooked orchestrator and is responsible for managing critical operations in the background, including processing button presses that trigger service worker logic. The code for this file is included in the background.js file and are detailed in Listing 6-2.

Listing 6-2. Configuring background.js with service worker logic

```
// Log a message indicating that the extension has been launched.
console.log("Extension launched...");

// Listen for the extension's action button click event.
chrome.action.onClicked.addListener(buttonClicked);

// Function to handle the button click event.
function buttonClicked(tab) {
  // Log a message indicating the background script is running when the
  button is clicked.
  console.log("Background script running...");
}

// Additional service worker logic can be added here
// This script operates separately from the extension and can perform
background tasks efficiently
// Developers can extend the functionality of the service worker to handle
various tasks, such as
// network requests, data management, and background processing.
```

6.2.3 Building the User Interface

The UI is constructed by combining HTML and CSS to create a popup consisting of a live graph, video quality metrics, a legend, and a button. These components are styled in the same HTML file—i.e., popup.html. This part of the program is responsible for providing an intuitive display of relevant metrics to the end user for video quality monitoring.

File Functionality

This section provides a detailed analysis of the popup.html file, outlining its specific functions as follows:

1. Specify the document structure consisting of the following:

    i. Document type and language.

    ii. Charset and title.

    iii. Modify the content security policy (CSP) to allow certain script sources.

2. Add references to external resources:

    i. Import the required font stylesheets from Google Fonts [5].

    ii. Import icons for the dashboard from the Font Awesome icon library [6].

    iii. Link a custom Bootstrap stylesheet downloaded from Bootstrap's official website [7].

3. Add internal CSS for modifying the component appearance:

    i. Set styling for the body, background, and buttons.

    ii. Define styling for user interface components, such as buttons, labels, and charts.

4. Display a title and chart area:

    i. Give the popup a title.

    ii. Create a container for embedding the chart area.

    iii. Embed a chart canvas with Chart.js.

5. Display the "Video Quality Performance" cards:

    i. Create a container for video quality metrics.

    ii. Display the latest MOS score, average MOS score, live quality rating, and dominant quality rating individually along with their icons.

6. Display the "Legend" cards:

    i. Create a container for the quality ratings.

    ii. Define the "Monitor" button.

7. Add downloaded and client scripts:

    i. Link the downloaded Chart.js library [4].

    ii. Link the custom pop.js script.

Libraries and Required Resources

Table 6-2 provides a concise overview of the stated libraries and downloadable scripts, together with their respective purposes.

Table 6-2. *Description of Libraries and Resources for the User Interface*

Library/ Resource	Purpose	Repository	Download Link	Location
Google Fonts	Provides custom fonts for the web page	[5]		popup.html
Font Awesome	Provides scalable vector icons	[6]		popup.html
bootstrap. min.css	Customized Bootstrap stylesheet for organizing row and columns	[7]	[8]	"css" folder inside project's root folder
chart.js	Provides an interactive chart for the live graph	[4]	[9]	Project's root folder
pop.js	Custom JS file for the logic	N/A		Project's root folder

Creating the Document Structure

Using the appropriate libraries and assets, the popup.html file is filled with the code provided in Listing 6-3. This code initiates the creation of the page structure. The location of the code for the remainder of the file is specified.

Listing 6-3. Creating the document structure

```
<!DOCTYPE html>
<html lang="en" dir="ltr">
   <head>
     <!-- Listing 6-4 is included here -->
     <style>
```

```
    <!-- Listing 6-6 is included here -->
    </style>
  </head>
  <body>
    <!-- Listing 6-5 is included here -->
  </body>
</html>
```

Adding the Document Details and References to External Resources

After establishing the foundation, the required fonts, icons, and styles are connected using the link tag and positioned inside the head element, as shown in Listing 6-4.

Listing 6-4. Adding document details and references to external resources

```
<meta charset="utf-8">
<!-- Set Content Security Policy to allow scripts from the same origin
('self') and unsafe-eval for script execution -->
<meta http-equiv="Content-Security-Policy" content="script-src 'self'
'unsafe-eval'; object-src 'self'">
<!-- Title for the tab -->
<title>Video Quality Assessment</title>
<!-- Font Stylesheet -->
<link rel="stylesheet" type="text/css" href="http://fonts.googleapis.com/cs
s?family=Roboto:regular,bold&subset=Latin">
<link rel="stylesheet" type="text/css" href="http://fonts.googleapis.com/
css?family=Fira Sans:regular,bold&subset=Latin">
<link rel="stylesheet" type="text/css" href="http://fonts.googleapis.com/
css?family=Roboto Mono:regular,bold&subset=Latin">
<!-- Icons Stylesheet -->
<link href="https://cdnjs.cloudflare.com/ajax/libs/font-
awesome/6.4.2/css/all.min.css" rel="stylesheet">
<!-- Customized Bootstrap Stylesheet -->
<!-- This file is downloaded from "https://getbootstrap.com/" online and
placed in the "css" folder. -->
<link href="css/bootstrap.min.css" rel="stylesheet">
```

263

Adding the Graph and Dashboard Components

The body tag houses the components that are seen by the user. These elements are coded as in Listing 6-5, along with their respective descriptions and purposes as explained in the comments. Within the body, the external scripts, i.e., chart.js and pop.js, are referenced. They are placed at the end to allow the complete loading of the document object model (DOM) prior to execution.

Listing 6-5. Adding the graph and dashboard components

```
<!-- Title to be displayed at the top of popup -->
<p id="title" class="mb-0">Video Quality Assessment</p>

<!-- Chart Area -->
<div class="container-fluid pt-2">
   <div class="row g-2">
      <div class="col-sm-12 col-xl-3">
         <!-- Container for the chart -->
         <div class="bg-secondary rounded d-flex align-items-center
         justify-content-between p-3">
            <!-- Canvas for the chart -->
            <canvas id="myChart" height="250"></canvas>
         </div>
      </div>
   </div>
</div>

<!-- Dashboard Components - Video Quality Performance Cards -->
<div class="dashboard-container">
   <!-- Title for the Video Quality Performance -->
   <p class="dashboard-title">Video Quality Performance</p>
   <!-- Row for the Video Quality Performance Cards -->
   <div class="dashboard-row">
      <!-- Card for Latest MOS Score -->
      <div class="dashboard-card">
         <!-- Icon for Latest MOS Score -->
         <i class="icon fas fa-video"></i>
```

```
<!-- Latest MOS Score Metric to be displayed -->
<span class="metric" id="results_mosScore" style="color:
#FF6384;">Hold on...</span>
<!-- Description for Latest MOS Score -->
<p class="description">Latest MOS Score</p>
</div>
<!-- Card for Average MOS Score -->
<div class="dashboard-card">
    <!-- Icon for Average MOS Score -->
    <i class="icon fas fa-vial"></i>
    <!-- Average MOS Score Metric to be displayed -->
    <span class="metric" id="results_averageMosScore" style="color:
    #2CD3E1;">Hold on...</span>
    <!-- Description for Average MOS Score -->
    <p class="description">Average MOS Score</p>
</div>
<!-- Card for Live Video Quality Rating -->
<div class="dashboard-card">
    <!-- Icon for Live Video Quality Rating -->
    <i class="icon fas fa-heart-pulse"></i>
    <!-- Live Video Quality Rating Metric to be displayed -->
    <span class="metric" id="results_liveQualityRating" style="color:
    #45FFCA;">Hold on...</span>
    <!-- Description for Live Video Quality Rating -->
    <p class="description">Live Quality Rating</p>
</div>
<!-- Card for Dominant Video Quality Rating -->
<div class="dashboard-card">
    <!-- Icon for Dominant Video Quality Rating -->
    <i class="icon fas fa-ranking-star"></i>
    <!-- Dominant Video Quality Rating Metric to be displayed -->
    <span class="metric" id="results_dominantQualityRating"
    style="color: #FF4DC0;">Hold on...</span>
    <!-- Description for Dominant Video Quality Rating -->
    <p class="description">Dominant Quality Rating</p>
```

```
            </div>
        </div>
</div>

<!-- Row for Legend and Monitor Button -->
<div class="row g-2" style="padding-top: 2px; padding-bottom: 2px;">
    <!-- Column for Legend and Monitor Button -->
    <div class="col-sm-12 col-xl-3">
        <!-- Dashboard Components - Legend and Monitor Button -->
        <div class="dashboard-container">
            <!-- Title for the Legend -->
            <p class="dashboard-title">Legend</p>
            <!-- Row for the Legend and Monitor Button -->
            <div class="dashboard-row">
                <!-- Card for Excellent Score -->
                <div class="dashboard-card" style="height: 55px;">
                    <!-- Excellent Score Legend to be displayed -->
                    <span class="metric" id="legend_excellent" style="color:
                    #DFCCFB;">0 - 20</span>
                    <!-- Description for Excellent Score -->
                    <p class="description">Excellent</p>
                </div>
                <!-- Card for Great Score -->
                <div class="dashboard-card" style="height: 55px;">
                    <!-- Great Score Legend to be displayed -->
                    <span class="metric" id="legend_great" style="color:
                    #DFCCFB;">20 - 40</span>
                    <!-- Description for Great Score -->
                    <p class="description">Great</p>
                </div>
                <!-- Card for Good Score -->
                <div class="dashboard-card" style="height: 55px;">
                    <!-- Good Score Legend to be displayed -->
                    <span class="metric" id="legend_good" style="color:
                    #DFCCFB;">40 - 65</span>
```

```
            <!-- Description for Good Score -->
            <p class="description">Good</p>
        </div>
        <!-- Card for Fair Score -->
        <div class="dashboard-card" style="height: 55px;">
            <!-- Fair Score Legend to be displayed -->
            <span class="metric" id="legend_fair" style="color:
            #DFCCFB;">65 - 80</span>
            <!-- Description for Fair Score -->
            <p class="description">Fair</p>
        </div>
        <!-- Card for Poor Score -->
        <div class="dashboard-card" style="height: 55px;">
            <!-- Poor Score Legend to be displayed -->
            <span class="metric" id="legend_poor" style="color:
            #DFCCFB;">80 - 90</span>
            <!-- Description for Poor Score -->
            <p class="description">Poor</p>
        </div>
        <!-- Card for Bad Score -->
        <div class="dashboard-card" style="height: 55px;">
            <!-- Bad Score Legend to be displayed -->
            <span class="metric" id="legend_bad" style="color:
            #DFCCFB;">> 90</span>
            <!-- Description for Bad Score -->
            <p class="description">Bad</p>
        </div>
        <!-- Button for starting the monitoring process, i.e.,
        triggering the server -->
        <button id="monitorButton">Monitor</button>
    </div>
  </div>
 </div>
</div>
```

```
<!-- Include scripts for chart.js and pop.js at the end of the body to
ensure DOM is fully loaded before execution -->
<script src="chart.js"></script>
<script src="pop.js"></script>
```

Styling the Components through Internal CSS

Listing 6-6 gives the code needed to add styling to the HTML components within the
style tag so as to improve the usability of the application.

Listing 6-6. Styling the components through internal CSS

```
/* Styles for the body */
body {
background-color: #131F33;
color: #92ABCF;
font-family: "Roboto";
font-size: 15px;
width: 710px;
}
/* Styles for the body's background */
.bg-secondary {
background-color: #0F1724 !important;
}
/* Styles for any primary text */
.text-primary {
color: #92ABCF !important;
}
/* Styles for the "Monitor" Button */
#monitorButton {
display: flex;
width: 200px;
height: 30px;
align-content: right;
margin-top: 16px;
margin-bottom: 10px;
margin-left: 10px;
```

```css
margin-right: 10px;
text-transform: uppercase;
flex-direction: row;
-moz-box-align: center;
align-items: center;
gap: 0.375rem;
border: medium none;
box-shadow: rgba(0, 0, 0, 0.2) 0px 3px 1px -2px, rgba(0, 0, 0, 0.14) 0px
2px 2px 0px, rgba(0, 0, 0, 0.12) 0px 1px 5px 0px;
font-weight: bold;
font-family: Fira Sans, Arial, Helvetica, sans-serif;
cursor: pointer;
transition: background-color 250ms cubic-bezier(0.4, 0, 0.2, 1) 0ms,
box-shadow 250ms cubic-bezier(0.4, 0, 0.2, 1) 0ms, border 250ms cubic-
bezier(0.4, 0, 0.2, 1) 0ms;
-moz-box-pack: center;
justify-content: center;
border-radius: 12px;
padding: 0.625rem 0.75rem;
font-size: 0.875rem;
color: rgb(0, 0, 0);
background-color: rgb(17, 236, 229);
}
/* Styles for appearance of button upon hover */
#monitorButton:hover {
background-color: #bd1515;
color: white;
transition: .5s;
}
/* Styles for the title of the popup */
#title {
padding-top: 2px;
padding-left: 15px;
text-align: left;
font-family: Roboto Mono;
```

```css
font-weight: bold;
font-size: 14px;
color: #DFCCFB;
}
/* Styles for the title of each dashboard row */
.dashboard-title {
color: #DFCCFB;
margin-left: 10px;
margin-bottom: -5px;
font-size: 13px;
font-family: 'Roboto Mono';
}
/* Styles for a dashboard row */
.dashboard-row {
display: flex;
justify-content: space-between;
}
/* Styles for a dashboard card */
.dashboard-card {
padding: 7px;
margin: 4px;
border-radius: 10px;
background-color: #101424;
position: relative;
width: 225px;
}
/* Styles for a dashboard container */
.dashboard-container {
background-color: #1d243f;
margin-top: 2px;
margin-left: 10px;
padding-left: 5px;
border-radius: 10px;
position: relative;
width: 97%;
}
```

270

```
/* Styles for a dashboard icon */
.icon {
position: absolute;
top: 10px;
right: 10px;
font-size: 25px;
color: #DFCCFB;
}
/* Styles for a dashboard metric */
.metric {
font-size: 18px;
}
/* Styles for a dashboard description */
.description {
font-size: 11px;
font-family: Fira Sans;
}
```

Adding the User Interface to Google Chrome

The steps outlined next are followed to add the browser extension frontend to Google Chrome:

1. On Google Chrome's main menu, the three dots icon is clicked on, before selecting "Extensions." Then, "Manage Extensions" is located and clicked on (Figure 6-5). This opens up Google Chrome Extensions.

Figure 6-5. *Accessing the extensions menu*

2. As depicted in Figure 6-6, the "Load Unpacked" option is selected.

Figure 6-6. *Loading an unpacked extension*

3. The project's root directory is then selected, as shown in Figure 6-7.

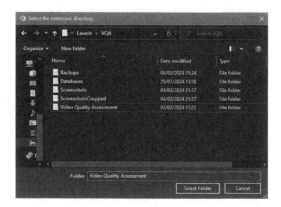

Figure 6-7. *Selecting the project's root directory*

4. At this point, the application is added to the Extensions menu (Figure 6-8).

Figure 6-8. *VQA application added to the Extensions menu*

5. The VQA application is then pinned at the top right of the browser for easy access, by clicking on the puzzle icon and the pin icon, as shown in Figure 6-9.

Figure 6-9. *Pinning the application for quick access*

Visualizing the User Interface

Figure 6-10 shows the completed user interface after clicking on the application in the quick access bar.

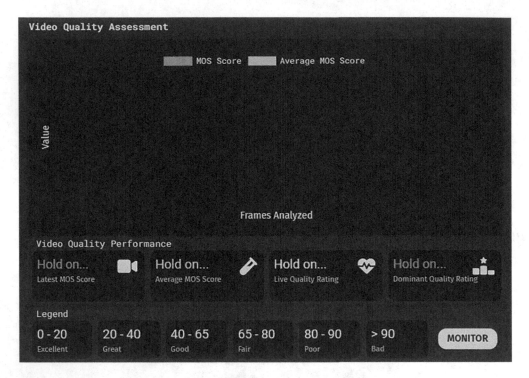

Figure 6-10. *VQA application completed user interface*

6.2.4 Building the Client Script

The client's functionality is implemented using JavaScript and is contained in a single file called pop.js. This file essentially manages communication with the server using WebSockets [2], refreshes the UI with the received metrics, and coordinates the real-time monitoring graph using Chart.js.

File Functionality

The following is a breakdown of the pop.js file:

1. Declare global variables for use throughout the script.

2. Query the Chrome tab using Chrome extension API:

    a. Retrieve information about the active and last focused-on tab.

    b. Retrieve the URL of the active tab.

3. Create a WebSocket connection to the server:

   a. Implement method for receiving messages from the server.

   b. Handle the connection using the open event.

   c. Parse and log messages.

4. Receive and handle messages from the server:

   a. Parse JSON messages containing real-time video metrics and graph updates.

   b. Update HTML elements with the video quality performance parameters.

   c. Update the Chart.js chart dynamically with the latest metrics.

5. Create an event listener for the "Monitor" button:

   a. Add an event listener to the button.

   b. Call a method that takes a screenshot of the current page view using the Chrome extension API.

   c. Calculate the number of 8 KB chunks it would take to send the image to the server.

   d. Send the number of chunks the server should expect, as metadata.

   e. Send each chunk to the server.

6. Create an event listener for the DOM:

   a. Add an event listener to the DOM.

   b. Create configurations for the two datasets and initialize the Chart.js chart object.

7. Implement a method to update the chart every second based on a specified label.

8. Implement a method to take a screenshot of the current page upon request from the Node.js server.

Creating the Script Structure

After importing and referencing the necessary libraries in the popup.html file, the pop.js script is filled by first defining the structure, which includes the following essential steps:

- Initializing the global variables
- Retrieving the URL of the currently active and previously focused-on tab
- Attaching an event listener to the DOM to trigger execution immediately upon page load
- Developing a procedure to dynamically modify the chart

Listing 6-7 defines the script structure, along with the placement of code for other functionalities.

Listing 6-7. Creating the script structure

```
// Declare the global objects that are called throughout the script.
var monitorButton; // Variable for the "Monitor" object.
var myChart; // Variable for the Chart.js object.

// Query the active and last focused-on tab to retrieve the current URL.
chrome.tabs.query({
    active: true,
    lastFocusedWindow: true,
},
    function(tabs) {

        <!-- Listing 6-9 is included here -->

    }
);

// Add an event listener on the DOM to execute as soon as the page loads.
document.addEventListener('DOMContentLoaded', function() {

    <!-- Listing 6-8 is included here -->

}, false);
```

```
// Method to update the chart after every frame is analyzed by the servlet.
function addData(chart, label, data, updateLabel) {

    <!-- Listing 6-10 is included here -->

}
```

Initializing and Configuring the Chart on Page Load

The DOMContentLoaded event triggers the execution of statements immediately after the HTML document's DOM has been completely loaded. The chart's area is referenced using the myChart variable, as indicated in popup.html. To accept the two separate datasets that are envisioned on the graph, the chart is initialized with the appropriate specifications. The following operations are executed by attaching an event listener to the whole document, which is promptly triggered when the page is loaded:

1. Get the DOM element for the "Monitor" button and assign it to the global variable monitorButton.

2. Create dataset configurations:

    i. Create two separate datasets for "MOS Score" and "Average MOS Score."

    ii. For each dataset, specify the label, background color, border color, border width, and point radius.

3. Create a data object that encapsulates all the dataset settings and is then fed directly as a parameter to Chart.js.

4. Create an options object that styles the appearance of the Chart.js chart, with settings including responsiveness, aspect ratio, line tension, axis configurations, tick display settings, and legend styles.

5. Create a config object for the Chart.js chart where the type of the chart, the data to be included, and the specified options are all encapsulated.

0. Create a new Chart.js instance using the myChart identifier, thus binding all the settings to the HTML element. This indicates the object where the line chart is rendered.

The code snippet for the specified functionalities is given in Listing 6-8.

Listing 6-8. Initializing and configuring the chart on page load

```
// Get the DOM element for the "Monitor" button.
monitorButton = document.getElementById('monitorButton');

// Create the dataset configuration for the "MOS Score" line chart.
const dataset_MOS = {
    label: 'MOS Score',
    backgroundColor: 'rgb(255, 99, 132)',
    borderColor: 'rgb(255, 99, 132)',
    borderWidth: 1.2,
    pointRadius: 1.6
};

// Create the dataset configuration for the "Average MOS Score" line chart.
const dataset_AvgMOS = {
    label: 'Average MOS Score',
    backgroundColor: 'rgb(44, 211, 225)',
    borderColor: 'rgb(44, 211, 225)',
    borderWidth: 1.2,
    pointRadius: 1.6
};

// Create the data object that encapsulates all datasets into one.
const data = {
    datasets: [dataset_MOS, dataset_AvgMOS]
};

// Create the options object, which is responsible for styling the graph's
appearance.
const options = {
    // Set the width to be responsive to the area allocated.
    responsive: true,
    // Disallow the aspect ratio to change.
    maintainAspectRatio: false,
    elements: {
        line: {
            tension: 0.35, // Add a slight curvature to the line graphs.
```

```
        },
    },
    // Modify x-axis configurations.
    scales: {
        x: {
            display: true, // Show x-axis.
            title: {
                display: true, // Show title of x-axis.
                text: 'Frames Analysed', // Change label of x-axis.
                color: '#DFCCFB', // Change color of x-axis label.
                font: {
                    family: 'Fira Sans', // Change font of x-axis label.
                    size: 14, // Change font size of x-axis label.
                },
            },
            ticks: {
                display: true, // Show ticks on x-axis.
                color: '#6b7e9b', // Change label color of x-axis tick.
                font: {
                    family: 'Ubuntu', // Change font of x-axis tick.
                    size: 10, // Change font size of x-axis tick.
                },
                stepSize: 10, // Change the step size for ticks on x-axis.
            },
        },
        y: {
            display: true, // Show y-axis.
            title: {
                display: true, // Show title of y-axis.
                text: 'Value', // Change label of y-axis.
                color: '#DFCCFB', // Change color of y-axis label.
                font: {
                    family: 'Fira Sans', // Change font of y-axis label.
                    size: 14, // Change font size of y-axis label.
                },
```

```javascript
                    },
                    ticks: {
                        display: false, // Disable ticks on y-axis.
                        color: '#6b7e9b', // Change label color of y-axis tick.
                        font: {
                            family: 'Ubuntu', // Change font of y-axis tick.
                            size: 13, // Change font size of y-axis tick.
                        },
                    },
                },
            },
            plugins: {
                legend: {
                    labels: {
                        color: '#DFCCFB', // Change legend label color.
                        font: {
                            family: 'Roboto Mono', // Change font of legend labels.
                        }
                    }
                }
            }
        }
    };

    // Configuration object for the line chart.
    const config = {
        type: 'line', // Set the chart type.
        data: data, // Set the data for the chart.
        options: options // Set the options for styling and customization.
    };

    // Create a new Chart instance using the configuration and attach it to the
    canvas element with the id "myChart".
    myChart = new Chart(document.getElementById('myChart'), config);
```

Real-Time WebSocket Communication and Dynamic Graph Updates

The main tasks of the client include establishing a connection with the server on a designated port before enabling the transmission of messages over the established connection. In addition, it has the capability to transmit messages on demand. However, due to the server's primary role in doing most of the computational tasks in this application, the client mostly focuses on receiving messages from the server, together with taking the successive screenshots. This is further elaborated as follows:

1. Get the URL of the active tab and store it in a variable.

2. Initialize a `time` variable that keeps track of the number of frames being analyzed.

3. Create constants to dynamically differentiate between the two graphs.

4. Establish a WebSocket connection on "ws://localhost:8081/", i.e., through port 8081, for real-time communication.

    i. To send a message, the `ws.open` event is used. In this case, the URL of the active tab stored earlier is sent to the server so it knows which site is being visited. This is crucial in manipulating the screenshot taken since it can be cropped according to the coordinates of the video frame. In this application, YouTube and Twitch are currently supported. However, the cropping coordinates for any video streaming website can be acquired and the code modified accordingly.

    ii. To receive a message, the `ws.onmessage` event is used. This method is automatically triggered upon receipt of any message from the server. The latter is programmed to request for another screenshot once it receives the results from the servlet, which it relays to the client. The client is thus coded to verify if the message from the server includes the "screenshot" keyword, as well as whether the number of frames analyzed is sixty.

    iii. The message from the server also contains the video quality metrics as generated by the servlet. These values are used to update the HTML elements and the graph.

5. Check the receipt of a message related to any error that occurred at the server side and generate an alert if so.

6. Trigger real-time graph updates using the addData() method, which takes as input the chart object, the timestep, the real-time value, and a tag for dictating which dataset is being populated among the two.

7. Create an event listener for the "Monitor" button:

    i. Call a method that takes a screenshot of the current page view using the Chrome extension API.

    ii. Implement the takeScreenshot() method outside of the "Monitor" button's event listener.

Listing 6-9 gives the code that coordinates the duplex communication with the server, contains the mechanism to take screenshots, and updates the graph in real time.

Listing 6-9. Real-time WebSocket communication and dynamic graph updates

```
// Store the URL of the active tab in a variable.
var tempurl = tabs[0].url;

// Variable for iterating time on the graph.
// It is used to keep track of the frames being analyzed.
var time = 0;

// Constants for updating graph data.
const updateMOS = "MOS";
const updateAverageMOS = "AverageMOS";

// Create a WebSocket connection to the server.
const ws = new WebSocket('ws://localhost:8081/');

// Handle the WebSocket connection "open" event.
ws.onopen = function() {
    // Log a message for when the extension has successfully connected to
    the server.
    console.log('WebSocket Client Connected');
```

```javascript
    // Send the URL of the active tab to the server for it to know which
    site is being visited.
    // Specify the type as "sitedata" and send the message in a
    JSON format.
    ws.send(JSON.stringify({type:'sitedata', value:tempurl}));
};

// Handle received messages from the server.
ws.onmessage = function(e) {
    // Log any received message.
    console.log("Received from server: " + e.data);

    // Parse the JSON message for real-time metrics and graph updates.
    // This occurs as soon as the NodeJS server receives a processed reply
    from the Java Servlet.
    const msg = JSON.parse(e.data);

    // Extract the message from the server requesting another screenshot.
    var screenshotMessage = msg.screenshot;

    // Extract the message from the server containing the MOS score.
    var mosScore = parseFloat(msg.videoScore);

    // Extract the message from the server containing the Average
    MOS score.
    var averageMosScore = parseFloat(msg.averageVideoScore);

    // Extract the message from the server containing the Video
    Quality Rating.
    var liveQualityRating = msg.videoRating;
    // Since this rating is live, update the HTML element with the received
    metrics from the server.
    document.getElementById("results_liveQualityRating").innerHTML =
    liveQualityRating;

    // Extract the message from the server containing the Video
    Quality Rating.
    var dominantQualityRating = msg.dominantRating;
```

```
// Since this rating is live, update the HTML element with the received
metrics from the server.
document.getElementById("results_dominantQualityRating").innerHTML =
dominantQualityRating;

// Add data to the chart using the addData() method by parsing the
following parameters:
// 1. Chart object.
// 2. Timestep (relative to when the result for a frame is being
received from the Java Servlet).
// 3. Real-time value.
// 4. Tag for guiding the variable to the right set of values on
the chart.
// Update the MOS graph with the information from the server.
addData(myChart, time, parseFloat(mosScore), updateMOS);
// Update the Average MOS graph with the information from the server.
addData(myChart, time, parseFloat(averageMosScore), updateAverageMOS);

// Handle messages related to any error sent by the server.
// The following statement checks whether the message received contains
the "error" keyword.
// This keyword is set at the server side.
if (e.data.includes("error")) {
    // Generate an alert to let the user know that video quality cannot
    be measured, along with the reason.
    alert("Unable to measure video quality. Reason: " + msg.error);
}

// Increment time for the next data point.
time = time + 1;

// If the number of frames analyzed is under 61.
if (e.data.includes("screenshot") && time < 61) {
    // Call the method to take a screenshot.
    takeScreenshot();
}
```

```
    // If the number of frames analyzed reaches 60.
    // Update the HTML elements with the received metrics from the server.
    if (time == 60) {
        document.getElementById("results_mosScore").innerHTML = mosScore;
        // Round the average latency to two decimal places using the
        toFixed() method.
        document.getElementById("results_averageMosScore").innerHTML =
        averageMosScore.toFixed(2);
    }
};

// Add an event listener to the "Monitor" button.
monitorButton.addEventListener('click', function() {
    // Call the method to take a screenshot.
    takeScreenshot();
}, false);

// Method to capture a screenshot of the current page.
function takeScreenshot() {
    // Capture a screenshot.
    chrome.tabs.captureVisibleTab({ format: "png" }, function
    (screenshotUrl) {
        // Create a Blob from the data URL.
        fetch(screenshotUrl)
        .then((res) => res.blob())
        .then((blob) => {
            // Create a Promise to handle the conversion of Blob to
            ArrayBuffer.
            return new Promise((resolve) => {
                // Create a FileReader instance.
                const reader = new FileReader();
                // Resolve with the result when the reader finishes
                loading.
                reader.onloadend = () => resolve(reader.result);
                // Read the Blob as ArrayBuffer.
                reader.readAsArrayBuffer(blob);
            });
```

```
    })
    .then((arrayBuffer) => {
        // Specify a chunk size in which the image is broken down to be
        sent to the server.
        const chunkSize = 8192; // e.g., 8 KB chunks

        // Calculate the number of chunks needed.
        const totalChunks = Math.ceil(arrayBuffer.byteLength /
        chunkSize);

        // Send metadata first (total number of chunks).
        ws.send(JSON.stringify({type: "metadata", totalChunks}));

        // Chunk and send the data into as many chunks required.
        // Iterate through the chunks and send them over the WebSocket
        connection.
        for (let i = 0; i < totalChunks; i++) {
            // Calculate the start and end positions for the
            current chunk.
            const start = i * chunkSize;
            const end = Math.min((i + 1) * chunkSize, arrayBuffer.
            byteLength);
            // Slice the arrayBuffer to obtain the current chunk.
            const chunk = arrayBuffer.slice(start, end);
            // Send the chunk over the WebSocket connection.
            ws.send(chunk);
        }
    })
    .catch((error) => {
        // Log any errors in capturing the screenshot.
        console.error("Error capturing screenshot:", error);
    });
    });
}
```

Real-Time Chart Update Mechanism

This section pertains to the addData() method, developed to refresh the Chart.js line graph at the intervals dictated by the VQA block. As shown in the previous listing, it is called once the results are obtained by the client. The method's inputs are as follows:

- The Chart.js object, i.e., myChart, which stays consistent across both datasets

- The label representing the frame index for which a result has been obtained.

- The real-time video metric obtained from the server.

- The previously defined tag, which specifies which dataset to update among the two.

Furthermore, when the initial set of numbers is updated on the graph, the y-axis ticks, which were previously disabled using the options object, are now enabled to provide a smooth and uninterrupted display. The code for the addData() function is given in Listing 6-10.

Listing 6-10. Real-time chart update mechanism

```
// Determine for which dataset the method has been called through the tag
previously made.
// If the tag represents the MOS.
if (updateLabel === "MOS") {
    // Push the label, i.e., the frame instant.
    chart.data.labels.push(label);
    // Push the data, i.e., the real-time metric.
    chart.data.datasets[0].data.push(data);
}
// If the tag represents the Average MOS.
else if (updateLabel === "AverageMOS") {
    // Push the data, i.e., the real-time metric.
    chart.data.datasets[1].data.push(data);
}
```

```
// Update the chart options to display the y-axis ticks when this method is
first called.
// This is done since the y-axis ticks are disabled at first.
myChart.options.scales.y.ticks.display = true;

// Update the chart.
chart.update();
```

6.3 Server Program Structure for VQA

The major responsibilities of the Node.js server for this application are given here:

1. **Import modules**: The file starts by adding modules using the require keyword. The libraries added help create the HTTP server, make the WebSocket connection, perform file system operations with privileges, process the screenshot, and handle the request-response mechanism to and from the servlet.

2. **Initialize global variables**: Declare and initialize the objects to be used throughout the script.

3. **Create an HTTP server**: An HTTP server is hosted on port 8081 and listens for incoming requests from a client requesting connection. This server hosts the WebSocket connection.

4. **Handle WebSocket connections**: The WebSocket server is created and bound to the HTTP server. It then listens for connections from a connecting client and obtains data such as the site name, number of chunks, and binary data, before triggering the VQA process.

5. **Process the screenshot**: The image is reconstructed by appending the 8 KB chunks received from the client and cropped using the sharp library to only consist of the video frame.

6. **Communicate with the Java servlet**: Using the axios library, HTTP POST requests are sent to the hosted Java servlet on localhost to calculate the MOS based on the image cropped and stored. The responses are handled and map to an interpretable video quality rating.

7. **Monitor video quality**: The metrics updates are sent to the client
    over the WebSocket connection, including MOS scores, quality
    ratings, and average scores.

At every checkpoint, the script logs statements to provide insights into the
monitoring process. This includes the server's current status, the messages sent
and received, and the responses obtained from the servlet. Figure 6-11 outlines the
components of the server side.

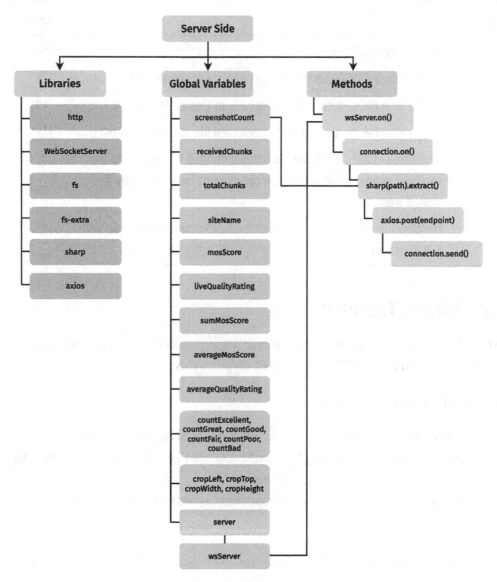

Figure 6-11. *Server-side program structure*

6.3.1 Libraries and Required Resources

The libraries are first installed locally using the npm install keyword in the command prompt. The installation instructions for each specific module are available in the download links in Table 6-3.

Table 6-3. *Description of Libraries and Resources for the Server*

Library/ Resource	Purpose	Download Link
http	Creates the HTTP server	[10]
websocket	Binds onto the HTTP server for the WebSocket connection	[2]
fs	Works as a file system, allowing access to files on the device	[11]
fs-extra	Works as a file system, allowing additional file system operations	[12]
sharp	Processes images easily by providing a method that can crop the received images based on preset coordinates determined by the visited website	[3]
axios	Handles the request-response architecture for communicating with the Java servlet	[13]

6.3.2 Adding Libraries

The specified libraries are added to the server's script using the require keyword at the beginning of the server.js file. Listing 6-11 displays this process.

Listing 6-11. Adding libraries

```
//Specify imports related to the server and monitoring process.
const http = require('http'); // Import the 'http' module for creating an
                             HTTP server.
const WebSocketServer = require('websocket').server; // Import the
                                                      WebSocket
                                                      server module.
const fs = require('fs'); // Import the 'fs' module for working with the
                          file system.
```

```
const fsExtra = require('fs-extra'); // Import the 'fs-extra' module for
                                       additional file system operations.
const sharp = require('sharp'); // Import the 'sharp' module for image
                                  processing.
const axios = require('axios'); // Import the 'axios' module for making
                                  HTTP requests.
```

6.3.3 Declaring Global Variables

Global variables play a crucial role in preserving state and facilitating data sharing across various sections of the script. Therefore, they are defined and initialized immediately after importing the libraries, as seen in Listing 6-12.

Listing 6-12. Declaring global variables

```
// Declare and initialize global variables used across the server's script.
// Variables related to image processing.
var screenshotCount = 1; // Counter for saving screenshots locally in
                            sequential order.
var receivedChunks = 0; // Counter for tracking received chunks of data.
var totalChunks; // Variable to store the total number of expected chunks.
var siteName = ""; // Variable to store the name of the website.

// Variables related to VQA.
var mosScore = 0; // Variable to store the Mean Opinion Score (MOS).
var liveQualityRating = ""; // Variable to store the live quality rating.
var sumMosScore = 0; // Variable to accumulate the sum of MOS scores.
var averageMosScore = 0; // Variable to store the average MOS score.
var averageQualityRating = ""; // Variable to store the average
                                  quality rating.

// Variables related to finding the dominant video quality rating.
var countExcellent = 0; // Counter for the number of occurrences of
                           "Excellent" rating.
var countGreat = 0; // Counter for the number of occurrences of
                       "Great" rating.
```

```
var countGood = 0; // Counter for the number of occurrences of
                        "Good" rating.
var countFair = 0; // Counter for the number of occurrences of
                        "Fair" rating.
var countPoor = 0; // Counter for the number of occurrences of
                        "Poor" rating.
var countBad = 0; // Counter for the number of occurrences of "Bad" rating.

// Variables that dictate which parts of the screen to crop depending on
the website being visited.
var cropLeft = 0; // Variable to store the left coordinate for
                        cropping images.
var cropTop = 0; // Variable to store the top coordinate for
                        cropping images.
var cropWidth = 0; // Variable to store the width for cropping images.
var cropHeight = 0; // Variable to store the height for cropping images.
```

6.3.4 Emptying the Screenshot Folders

This section involves emptying the folders where the screenshots are stored. It ensures that the folders are clean and ready to store new screenshots before the server starts listening for connections. Two folders, namely "Screenshots" and "ScreenshotsCropped," are first created inside the project's root directory. The former is coded to store raw screenshots of the web browser, while the latter stores the cropped screenshots. Since the fs-extra library can perform advanced file system operations, is it advised to be extremely careful when inserting the path to the folders to be emptied. The code for this part is given in Listing 6-13.

Listing 6-13. Emptying the screenshot folders

```
// Empty the "Screenshots" and "Screenshots-Cropped" folders.
// Please be careful which directory is being inserted here!
fsExtra.emptyDirSync("C://Users//Lavesh//VQA//Screenshots");
fsExtra.emptyDirSync("C://Users//Lavesh//VQA//ScreenshotsCropped");
```

6.3.5 Creating a WebSocket Server

As mentioned in Section 5.3.5, a WebSocket server is used in situations where achieving minimal delay is crucial. It enhances performance and responsiveness, allowing dashboards to be populated in real time. Listing 6-14 gives the code for creating an HTTP server and binding a WebSocket server to it.

Listing 6-14. Creating a WebSocket server

```
// Host a server connection on port 8081.
const server = http.createServer(); // Create an HTTP server instance.
server.listen(8081); // Listen on port 8081 for incoming HTTP requests.

console.log("Server started on port 8081..."); // Log a message indicating
                                                 that the server has started.
console.log("Waiting for connection..."); // Log a message indicating
                                             that the server is waiting for
                                             connections.

// Create a WebSocket server instance and attach it to the HTTP server
previously created.
const wsServer = new WebSocketServer({
    httpServer: server,
});
```

6.3.6 Listening for a Client Connection Request

This section encompasses several procedures that relate to the relationship between the client and the server, as indicated in Figure 6-3. The following explanations are provided:

- Listen for a connection request made by the client.

- Accept the connection.

- Obtain the metadata containing the number of chunks that the server should expect, as well as the site data variable containing the URL of the website being visited. These variables are sent by the client through JSON-formatted strings.

- Receive binary image data.

- Append the received image chunks to reconstruct the complete image.

- Store the original screenshot in PNG format in the "Screenshots" folder.

- Crop the original screenshot to obtain the video frame only, based on preset coordinates, as determined by the site name obtained.

- Store the cropped screenshot in PNG format in the "ScreenshotsCropped" folder.

- Make HTTP POST requests to the Java servlet.

- Obtain the MOS score response.

- Convert the MOS score into an interpretable video quality rating.

- Calculate the average MOS score based on the number of frames analyzed.

- Calculate the dominant video quality rating, i.e., the rating that occurs the most.

- Send the results back to the client through WebSockets.

Listing 6-15 provides the framework for the procedures associated with the activity of waiting for and approving a client request. The connection.on code block is now void of any content and is populated in the subsequent section.

Listing 6-15. Listening for a client connection request

```
// Listen for a request for connection by a client through the
WebSocket server.
wsServer.on("request", function (request) {
    // Accept a connection from a client.
    const connection = request.accept(null, request.origin);

    // Use the "message" keyword to accept incoming messages from
    the client.
    connection.on("message", function (message) {
    <!-- Listing 6-16 is included here -->
    });
```

```javascript
    // Use the "close" keyword to handle client disconnection.
    connection.on("close", function (reasonCode, description) {
        // Log a message to indicate that the client has disconnected.
        console.log("Client has disconnected.");
    });
});
```

Receiving Metadata and Site Data

With the connection.on code block being triggered upon any message being received by the server, it is first logged on the console. The metadata and sitedata variables contain the number of chunks from which the image is concatenated by the client and the URL of the website being browsed, respectively. These objects are purposely encapsulated in UTF-8 format. This contrasts with the actual image sent by the client, which is in binary format. This mechanism allows the WebSocket connection to differentiate between the metadata and sitedata variables, and the actual chunks of the screenshot. These checks are performed through a series of if statements, which identify which one is being received, before storing them in their respective global variables. Moreover, once the sitedata variable is obtained, the cropping coordinates are set depending on the website. These functions are performed as shown in Listing 6-16.

Listing 6-16. Getting messages from the client and creating a repeating function

```javascript
// Check if the received message type is in UTF-8 format.
if (message.type === "utf8") {
    // Parse the UTF-8 data from the received message.
    const messageReceived = JSON.parse(message.utf8Data);
    // Check the type of the received message (metadata or sitedata).
    if (messageReceived.type === 'metadata') {
        // Update the totalChunks variable with the received metadata.
        totalChunks = messageReceived.totalChunks;
        // Log the received metadata information.
        console.log('Received metadata:', totalChunks, 'chunks expected.');
    } else if (messageReceived.type === 'sitedata') {
        // Update the siteName variable with the received value.
        siteName = messageReceived.value;
        // Log the received sitedata information.
```

```
        console.log('Received sitedata:', siteName, 'is the link.');
        // Check the website link to set cropping coordinates accordingly.
        // For YouTube and Twitch, the coordinates have been set.
        // For other websites, the whole page is taken as the screenshot
        for the VQA block.
        if (siteName.includes("youtube")){
            // For videos on https://www.youtube.com/
            cropLeft = 98;
            cropTop = 80;
            cropWidth = 1280;
            cropHeight = 720;
        } else if (siteName.includes("twitch")) {
            // For videos on https://www.twitch.tv/
            cropLeft = 239;
            cropTop = 52;
            cropWidth = 1340;
            cropHeight = 752;
        } else {
            // For videos on any other websites (no cropping).
            cropLeft = 0;
            cropTop = 0;
            cropWidth = 1920;
            cropHeight = 953;
        }
    }
} else if (message.type === "binary") {
<!-- Listing 6-17 is included here -->
}
```

Reconstructing and Cropping the Image

This section details the process of receiving the binary image data from the client, reconstructing the whole image, and then cropping it to focus on the relevant content. The main actions are as follows, and are given in Listing 6-17.

- Get the binary data in an individual chunk.

- Convert the data received into a buffer.

- Append the chunk to a file and store it locally by specifying a path and the file's extension, i.e., ".png".

- Log the current number of chunks received out of the total.

- Check if all chunks have been received by comparing the number received with the total previously obtained through the metadata object.

- If so, specify the paths of the image to be cropped, as well as that of the destination, i.e., cropped image.

- Crop the original screenshot using the sharp library, with the preset coordinates obtained from the sitedata object.

- Send any errors obtained to the client or log a message to indicate the success of the operation.

Listing 6-17. Reconstructing and cropping the image

```
// If the message received is of type binary.
// Increment the counter for the number of chunks received.
receivedChunks++;

// Handle the binary data received.
const arrayBuffer = message.binaryData;

// Convert arrayBuffer to Buffer.
const buffer = Buffer.from(arrayBuffer);

// Append the chunk to a file (e.g., screenshot-1.png).
// For one screenshot, each of its chunks is appended to the same file to
create the final screenshot.
fs.appendFileSync("C://Users//Lavesh//VQA//Screenshots//screenshot-" +
screenshotCount + ".png", buffer);

// Log a message to indicate the index of chunks received.
console.log('Received chunk:', receivedChunks, 'of', totalChunks);
```

```
// If all chunks received, finalize and save the image file.
if (receivedChunks === totalChunks) {
    // This denotes that a full screenshot has been received from the
    client side.
    console.log("All chunks received. Screenshot saved.");

    // Specify the paths of the file before and after cropping.
    var pathBeforeCropping = "C://Users//Lavesh//VQA//Screenshots//
    screenshot-" + screenshotCount + ".png";
    var pathAfterCropping = "C://Users//Lavesh//VQA//ScreenshotsCropped//
    screenshot-" + screenshotCount + ".png";

    // Use the "Sharp" library to crop the image according to the preset
    coordinates.
    sharp(pathBeforeCropping)
      .extract({ left: cropLeft, top: cropTop, width: cropWidth, height:
      cropHeight })
      .toFile(pathAfterCropping, (err, info) => {
        if (err) {
            // Log an error message to indicate that cropping cannot be
            performed.
            console.error('Error cropping image:', err);
            // Create an error message to be sent over to the client.
            const errorMessage = {
                error: "Error cropping image."
            };
            // Encapsulate the error message in a JSON-
            formatted String and send it over the WebSocket connection.
            connection.send(JSON.stringify(errorMessage));
        } else {
            // Log a message to indicate that the image has been cropped
            successfully.
            console.log('Image cropped successfully:', info);
            <!-- Listing 6-18 is included here -->
```

```
    }
  });
// Set the number of received chunks to zero, as a reset for the next
screenshot.
receivedChunks = 0;
}
```

Querying the Servlet and Handling the Response

In this section, the process through which the server contacts the Java servlet for information to obtain the video quality MOS is detailed. It covers making HTTP POST requests, handling responses, mapping MOS scores to quality ratings, and calculating average scores, as detailed here:

- Make an HTTP POST request to the servlet's endpoint using the `axios` library.

- Specify the screenshot's index as a parameter.

- Handle the response by receiving a JSON response and storing it in a variable.

- Map the MOS score obtained to its quality rating for ease of understanding.

- Increment different counters to identify the most occurring quality rating.

- Calculate the average MOS score.

- Encapsulate the MOS score, average MOS score, video quality rating, dominant video quality rating, and a tag that requests the client for another screenshot into a JSON-formatted string.

- Send this variable over the WebSocket connection to the client using the `connection.send()` method.

- Log any errors obtained and send to the client.

These processes are demonstrated in Listing 6-18.

Listing 6-18. Querying the servlet and handling the response

```
// At this point, the cropped image containing the feed that the VQA block
needs is available.
// Log a message indicating that a request is being made to the Java HTTP
Apache Server (servlet) to return the MOS score.
console.log("Making request to servlet.")
// Use the "Axios" library to make an HTTP POST request to the servlet.
axios.post("http://localhost:8080/VQA/Score", null, {
    // Specify the parameter required by the servlet, in this case the
    screenshot count.
    params: {
        number: screenshotCount,
    },
})
.then((response) => {
    // Handle the response from the servlet.
    // Retrieve the MOS score from the JSON response.
    mosScore = response.data.score;
    mosScore = parseFloat(mosScore);
    // Log a message to indicate the MOS score obtained.
    console.log("Score obtained from servlet: " + mosScore);

    // Account for any abnormalities in case of extremely bad quality.
    if (mosScore > 100) {
        mosScore = 100;
    }
    // Map the MOS score obtained to its quality rating for ease of
    understanding.
    // Increment the respective counter based on the quality assigned.
    // This is to identify the most occuring video quality, i.e., the
    "Dominant Quality Rating."
    if (mosScore >= 0 && mosScore < 20) {
        liveQualityRating = "Excellent";
        countExcellent = countExcellent + 1;
    }
```

```
else if (mosScore >= 20 && mosScore < 40) {
    liveQualityRating = "Great";
    countGreat = countGreat + 1;
}
else if (mosScore >= 40 && mosScore < 65) {
    liveQualityRating = "Good";
    countGood = countGood + 1;
}
else if (mosScore >= 65 && mosScore < 80) {
    liveQualityRating = "Fair";
    countFair = countFair + 1;
}
else if (mosScore >= 80 && mosScore < 90) {
    liveQualityRating = "Poor";
    countPoor = countPoor + 1;
}
else if (mosScore >= 90) {
    liveQualityRating = "Bad";
    countBad = countBad + 1;
}
// Log a message to indicate the live quality rating.
console.log("Video Quality Rating: " + liveQualityRating);

// Create an array to store the counts for each quality rating.
var countQuality = [countExcellent, countGreat, countGood, countFair,
countPoor, countBad];
// Initialize variables to find the maximum count and its index.
var maxCount = countQuality[0]; // Assume the first one is the maximum.
var maxIndex = 0;
// Iterate through the countQuality array to find the maximum count and
its index.
for (var i = 1; i < countQuality.length; i++) {
    // Check if the current count is greater than the maximum count.
```

```javascript
            if (countQuality[i] > maxCount) {
                // Update the maximum count and its index.
                maxCount = countQuality[i];
                maxIndex = i;
            }
        }
    }
    // Create an array of labels representing the quality ratings.
    var labels = ['Excellent', 'Great', 'Good', 'Fair', 'Poor', 'Bad'];
    // Determine the dominant quality rating based on the maximum index.
    var dominantQualityRating = labels[maxIndex];

    // Log the dominant quality rating and its count to the console.
    console.log('The highest count label is:', dominantQualityRating);
    console.log('The highest count value is:', maxCount);

    // Calculate and log the Average MOS Score.
    sumMosScore = sumMosScore + mosScore;
    averageMosScore = sumMosScore / screenshotCount;
    console.log("Average MOS Score: " +averageMosScore);

    // Send a message to the client to ask for another screenshot.
    // Create a constant variable holding the scores and ratings measured/
    calculated.
    const screenshotUpdate = {
        screenshot: "sendScreenshot",
        videoScore: mosScore,
        averageVideoScore: averageMosScore,
        videoRating: liveQualityRating,
        dominantRating: dominantQualityRating
    };

    // Encapsulate the metrics message in a JSON-
    formatted String and send it over the WebSocket connection.
    connection.send(JSON.stringify(screenshotUpdate));

    // Increment the screenshot count to move to the next one.
    screenshotCount = screenshotCount + 1;
})
```

```
.catch((error) => {
    // Log an error message to indicate that the screenshot cannot be sent
    to the servlet.
    console.error('Error sending screenshot to servlet:', error);
    // Create an error message to be sent over to the client.
    const errorMessage = {
        error: "Error sending screenshot to servlet."
    };
    // Encapsulate the error message in a JSON-
    formatted String and send it over the WebSocket connection.
    connection.send(JSON.stringify(errorMessage));
});
```

6.4 Servlet Program Structure for VQA

In this section, the Java servlet, which is built in the Eclipse IDE and hosted locally using Apache Tomcat, is detailed. The major functions of the servlet are as follows:

1. **Handle the HTTP POST request**: The servlet is contacted through an HTTP POST request. When created, the Java file already consists of implementations for receiving a client request. The servlet is then responsible for successfully receiving the parameter sent in the request before launching its other tasks.

2. **Locate the image**: After receiving the screenshot number of the image to undergo quality assessment, the servlet dynamically looks for that file through its path.

3. **Introduce distortion to the image**: An array of similarity scores is produced through the comparison of the original image with its multiple pseudo reference images (MPRIs).

4. **Extract the local binary pattern (LBP) features**: After obtaining the distorted image, the binarization procedure is performed using the resulting differences in brightness.

5. **Predict the similarity index and MOS quality score**: The scores for blocking, ringing, blurring, and noising artifacts are stored in a twenty-dimensional feature vector before scaling the regressor. The LIBSVM library is used in this step.

Figure 6-12 illustrates the general program structure for the servlet.

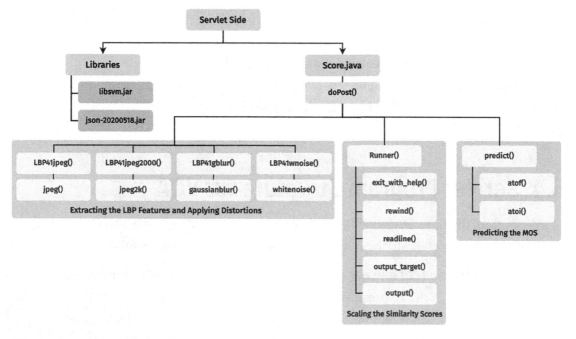

Figure 6-12. *Servlet-side program structure*

Moreover, the unified modeling language (UML) diagram generated by Eclipse and containing the global variables and methods is given in Figure 6-13.

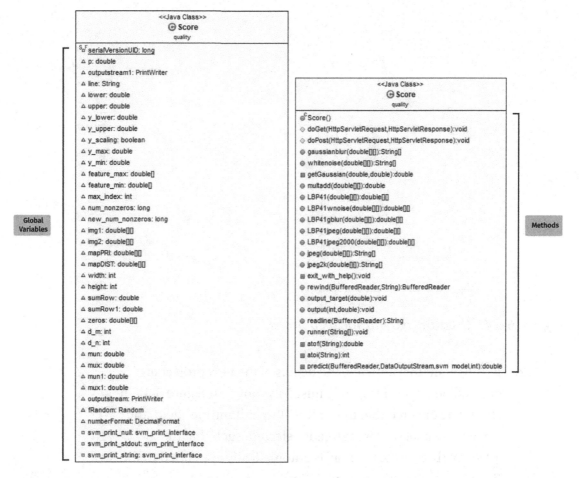

Figure 6-13. *UML diagram for Java servlet*

6.4.1 Creating a Java Servlet in Eclipse

To first create the project in Eclipse, the steps described are followed:

1. Eclipse is launched before clicking on "File" ➤ "New" ➤
    "Dynamic Web Project" (Figure 6-14).

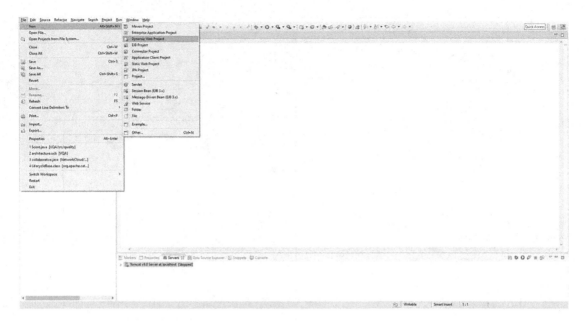

Figure 6-14. *Creating a dynamic web project*

2. In the menu that appears, the details of the new project are
   entered before clicking on "Finish," as shown in Figure 6-15. It
   should be ensured that under the "Target Runtime" field, the
   Apache Tomcat server is already selected, such that the
   `@WebServlet("/Score")` tag is automatically appended to the Java
   file upon project creation by Eclipse.

Figure 6-15. *Entering the project details*

3. To create the servlet file, which is a Java file, the created project on
    the left-hand side panel is right-clicked, before selecting "New"
    followed by "Servlet." This is shown in Figure 6-16.

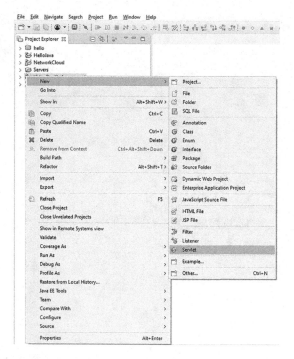

Figure 6-16. *Creating a servlet file*

4. The file is named Score.java, as shown by the "Class name"
   label, and located under the "quality" package, as depicted in
   Figure 6-17.

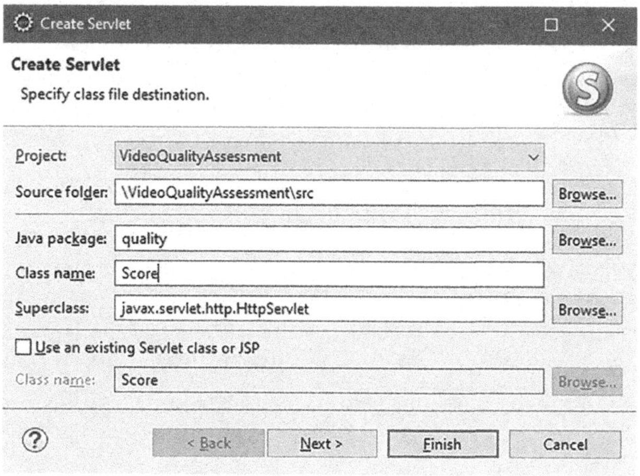

Figure 6-17. *Entering the file details*

5. Once created, the constructor, doGet(), and doPost() methods are automatically generated, as shown in Figure 6-18.

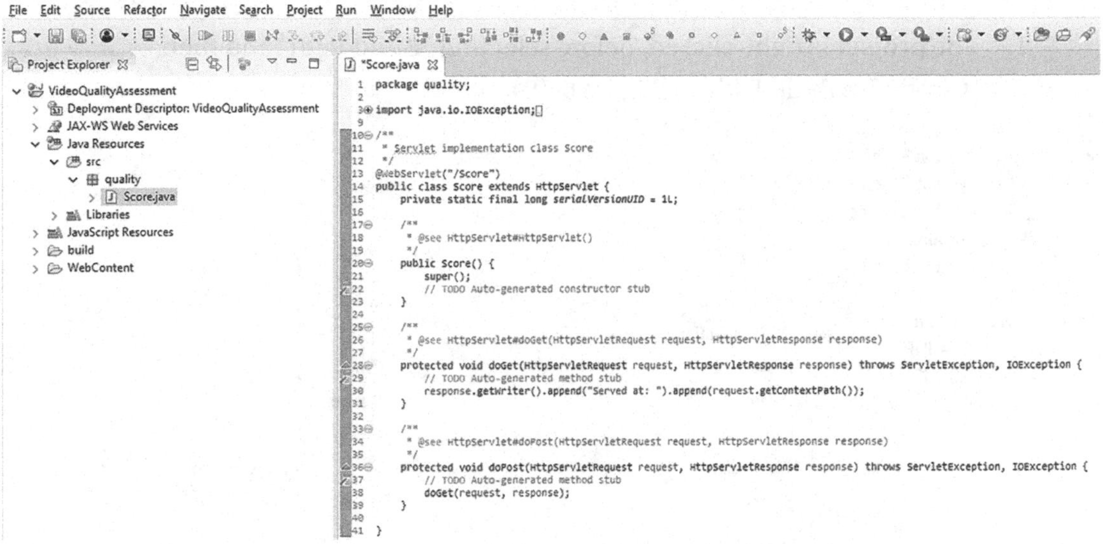

Figure 6-18. *Automatic generation of the methods in the servlet file*

This marks the creation of the servlet and the Java file required to host it locally.

6.4.2 Libraries and Required Resources

For this project, the libraries required are in the form of ".jar" files, as summarized in Table 6-4.

Table 6-4. *Description of Libraries and Resources for the Servlet*

Library/Resource	Purpose	Download Link
libsvm.jar	Provides functionalities to train SVM models, make predictions, and handle several parameters related to SVM algorithms	[14]
json-20200518.jar	Provides support for JSON (JavaScript Object Notation) data interchange format in Java. This notation is typically used to exchange data between a server and web applications given its lightweight nature.	[15]

6.4.3 Adding Libraries

The downloaded ".jar" files are then added to the project through the following steps:

1. The project is right-clicked, before selecting "Build Path" and then "Configure Build Path..." (Figure 6-19).

Figure 6-19. *Configuring the build path*

2. Next, the "Add External JARs..." option is selected, as shown in Figure 6-20.

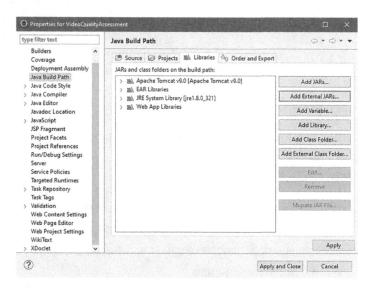

Figure 6-20. *Adding the external JARs*

3. The location of the files is browsed before selecting each of them (Figure 6-21).

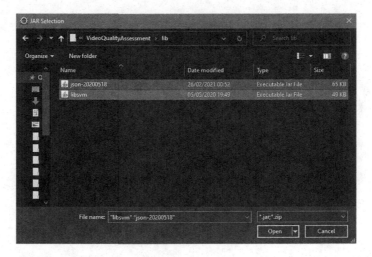

Figure 6-21. *Selecting the external JARs*

4. Additionally, the "WEB-INF" folder is located in the project's directory. It contains the "lib" folder, in which the ".jar" files are also pasted, as illustrated in Figure 6-22.

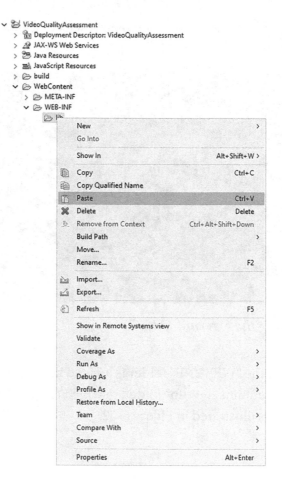

Figure 6-22. *Pasting the JARs in the lib folder*

6.4.4 Adding Imports

The imports pertaining to this project are first added to the Score.java file, as given in
Listing 6-19.

Listing 6-19. Adding imports

```
// Imports for input/output processing.
import java.awt.image.BufferedImage;
import java.io.BufferedReader;
import java.io.File;
import java.io.IOException;
import java.io.PrintWriter;
```

```
import java.io.BufferedOutputStream;
import java.io.DataOutputStream;
import java.io.FileNotFoundException;
import java.io.FileOutputStream;
import java.io.FileReader;

// Imports for handling the servlet.
import javax.imageio.ImageIO;
import javax.servlet.Servlet;
import javax.servlet.ServletException;
import javax.servlet.annotation.WebServlet;
import javax.servlet.http.HttpServlet;
import javax.servlet.http.HttpServletRequest;
import javax.servlet.http.HttpServletResponse;

// Imports for the JSON formatting.
import org.json.JSONObject;

// Imports for LIBSVM.
import libsvm.svm;
import libsvm.svm_model;
import libsvm.svm_node;
import libsvm.svm_parameter;
import libsvm.svm_print_interface;

// Other imports.
import java.util.Arrays;
import java.util.StringTokenizer;
import java.text.DecimalFormat;
import java.util.Random;
```

6.4.5 Declaring Global Variables

The variables used throughout the servlet are declared and initialized inside of the public Score class, as provided in Listing 6-20.

Listing 6-20. Declaring global variables

```
// Declare and initialize the global variables to be used throughout the
servlet.

double p = 0; // Prediction value, i.e., the MOS Score.

PrintWriter outputstream1; // Output stream for writing results.

String line = null; // Current line being processed.

// Parameters for scaling.
double lower = 0.0; // Lower bound for scaling.
double upper = 1.0; // Upper bound for scaling.
double y_lower; // Lower bound for y scaling.
double y_upper; // Upper bound for y scaling.
boolean y_scaling = false; // Flag indicating whether y scaling is applied.
double y_max = -Double.MAX_VALUE; // Maximum target value for y scaling.
double y_min = Double.MAX_VALUE; // Minimum target value for y scaling.
double[] feature_max; // Maximum values for each feature.
double[] feature_min; // Minimum values for each feature.
int max_index; // Maximum index of attributes.
long num_nonzeros = 0; // Number of non-zero elements.
long new_num_nonzeros = 0; // New number of non-zero elements.

// Arrays for image data.
double[][] img1;
double[][] img2;

// Maps for PRI and DIST.
double[][] mapPRI;
double[][] mapDIST;

// Width and height of the images.
int width, height;

// Sums of rows for image processing.
double sumRow, sumRow1;
```

```
// 2D array for zeros.
double[][] zeros;

// Dimensions of matrices (d_m rows, d_n columns).
int d_m, d_n;

// Parameters for normalization of image values.
double mun = 0.4305;
double mux = 0.7500;
double mun1 = 0.3350;
double mux1 = 0.3410;

// Output stream for general results.
PrintWriter outputstream;

// Random number generator.
Random fRandom = new Random();

// Decimal number format for output (to 2 decimal places).
DecimalFormat numberFormat = new DecimalFormat("#0.00");
```

6.4.6 Handling an HTTP POST Request from a Client

This section entails the development of the doPost() method, which is triggered when the servlet receives a request from a client. It thus coordinates the series of processing steps illustrated in Figure 6-23.

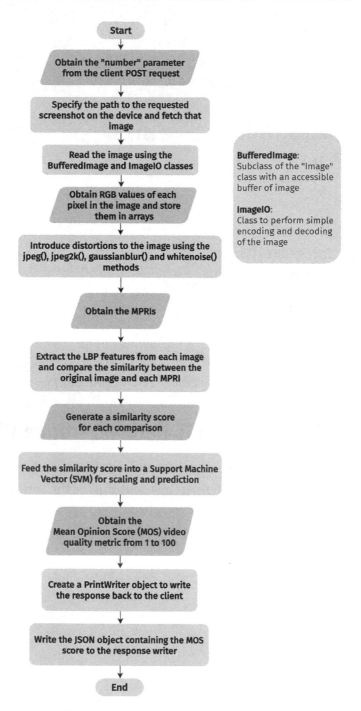

Figure 6-23. *Functionalities of the servlet*

The following is a breakdown of the functionalities of the doPost() method:

1. Get the "number" parameter from the HTTP request sent by the client.

2. Process the image read:

    a. Read the image file based on the "number" parameter using the read() method.

    b. Extract RGB values of each pixel in the screenshot using the getRGB() method, which converts the image to grayscale. Each red, green, and blue pixel value is averaged within the range 0–255 and stored in the img1 array.

    c. Extract the LBP features from the grayscale screenshot.

    d. Apply the compression and distortion techniques through the jpeg(), jpeg2000(), gblur(), and wnoise() methods.

3. Create the "dataset1.txt" and "dataset.txt" files for SVM input data and write the processed image data to these files.

4. Call the image processing runner:

    a. Define command-line arguments for image processing.

    b. Run the process using the runner() method.

5. Define command-line arguments for SVM scaling and prediction.

6. Set options such as prediction probability and output stream.

7. Load the SVM model from the specified file and use it to predict the MOS for the processed image data.

8. Obtain the MOS score as the prediction result.

9. Create a JSON object to store the MOS score and set the content type of the response as JSON.

10. Use a PrintWriter object to write the response back to the client.

The doPost() method is thus filled, as shown in Listing 6-21. The paths to the respective files should be changed accordingly.

Listing 6-21. Handling the HTTP POST request from a client

```
// Define the doPost method to handle HTTP POST requests.
protected void doPost(HttpServletRequest request, HttpServletResponse
response) throws ServletException, IOException {
        // Get the parameter sent by the client from the request made.
        // Use the "request" object to get the "number" parameter.
        String number = request.getParameter("number");

        // Use the "request" object to create the object to write the
        response to the client.
        PrintWriter writer = response.getWriter();

            // Create a try and catch block.
        try {
            try {
                // Read the image file specified by the client
                  through the "number" parameter received.
                BufferedImage image = ImageIO.read(new File("C:\\
                Users\\Lavesh\\VQA\\ScreenshotsCropped\\screenshot-" +
                number + ".png"));
                // Log the image being analysed.
                System.out.println("Image being analysed: screenshot-"
                + number + ".png");
                // Get the width of the image.
                int width = image.getWidth();
                // Get the height of the image.
                int height = image.getHeight();
                // Log the resolution of the image.
                System.out.println("Screenshot Resolution: " + width +
                " x " + height);

                int r; // Variable for Red.
                int g; // Variable for Green.
                int b; // Variable for Blue.
                // Initialize an array for extracting the value of
                each pixel.
                img1 = new double[width][height];
```

```java
        // Extract RGB values of each pixel in the image.
        for (int i = 0; i < width; i++) {
            for (int j = 0; j < height; j++) {
                // Store the extracted values in one
                2D array.
                img1[i][j] = image.getRGB(i, j) & 0xFF;
            }
        }
        // Extract RGB values of each pixel in the image.
        img2 = new double[width][height];
        for (int i = 0; i < width; i++) {
            for (int j = 0; j < height; j++) {
                // Store the extracted values in a second
                2D array.
                img2[i][j] = img1[i][j] / 255;
            }
        }
} catch (IOException ex) {
        // Catch and log any errors in reading the image.
        System.err.println("Could not read image: " + ex);
}

// Create the path for the "dataset1.txt" file.
String filename = "C:\\Users\\Lavesh\\Desktop\\FogFiles\\
dataset1.txt";
// Use a PrintWriter object to create the dataset1.
txt file.
outputstream = new PrintWriter(filename);
// Print a number to check if file is being written in.
outputstream.print("1 ");

// Apply the different image processing techniques one
after the other and write the results to a file.
jpeg(img1);
whitenoise(img1);
gaussianblur(img1);
jpeg2k(img1);
```

319

```
// Close the PrintWriter stream.
outputstream.close();

// Create the path for the "dataset.txt" file.
String filename1 = "C:\\Users\\Lavesh\\Desktop\\FogFiles\\
dataset.txt";

// Use another PrintWriter object to create the dataset.
txt file.
try {
    outputstream1 = new PrintWriter(filename1);
} catch (FileNotFoundException ex) {
    // Catch and log any errors in creating the file.
    System.out.println("File not found exception.");
}
// Print a number to check if file is being written in.
outputstream1.print("1 ");

// Fetch the file that defines the command-line arguments
for image processing.
String[] argv1 = {"C:\\Users\\Lavesh\\Desktop\\FogFiles\\
test_ind_scaled"};
try {
    // ** Run the image processing with the specified
    arguments.
    runner(argv1);
} catch (IOException ex) {
    // Catch and log any errors in handling the file.
    System.out.println("IO exception.");
}
// Close the PrintWriter stream.
outputstream1.close();

// Create variables for SVM scaling and prediction.
int i = 0; // Index for the for loop.
int predict_probability = 0; // Variable for the prediction
probability.
svm_print_string = svm_print_stdout; // **
```

```
// Fetch the files previously created that define the
command-line arguments for image processing.
String[] argv2 = {"C:\\Users\\Lavesh\\Desktop\\FogFiles\\
dataset.txt",
    "C:\\Users\\Lavesh\\Desktop\\FogFiles\\model",
    "C:\\Users\\Lavesh\\Desktop\\FogFiles\\output.txt"};

// Loop through the command-line arguments.
for (i = 0; i < argv2.length; i++) {
    // Check if the argument does not start with '-'.
    if (argv2[i].charAt(0) != '-') {
        // Exit the loop if the argument is not
        an option.
        break;
    }
    // Increment the index to move to the next argument.
    ++i;
    // Create a Switch statement on the second character of
    the current option.
    switch (argv2[i - 1].charAt(1)) {
        // If the option is 'b' (predict_probability).
        case 'b':
            // Convert the next argument to an integer
            and assign it to predict_probability.
            predict_probability = atoi(argv2[i]);
            break;
        // If the option is 'q' (svm_print_string).
        case 'q':
            // Set svm_print_string to svm_print_null and
            decrement i to process the next argument.
            svm_print_string = svm_print_null;
            i--;
            break;
```

```
            // If the option is not recognized.
            default:
                // Print an error message to standard error.
                System.err.print("Unknown option: " +
                argv2[i - 1] + "\n");
                // Exit the program with a help message.
                exit_with_help();
        }
    }

    // Check if there are enough arguments left after
    processing options.
    if (i >= argv2.length - 2) {
        // Exit the program with a help message.
        exit_with_help();
    }

    // // Create a try and catch block to handle potential
    exceptions.
    try {
        // Open a BufferedReader to read from the file
        specified in the command-line argument.
        BufferedReader input = new BufferedReader(new
        FileReader(argv2[i]));
        // Open a DataOutputStream to write to the file
        specified in the command-line argument (i + 2).
        DataOutputStream output = new DataOutputStream(new
        BufferedOutputStream(new FileOutputStream(argv2[i
        + 2])));
        // Load the SVM model from the file specified in the
        command-line argument (i + 1).
        svm_model model = svm.svm_load_model(argv2[i + 1]);
        // Check if predict_probability is set to 1.
        if (predict_probability == 1) {
            // Check if the loaded model supports probability
            estimates.
```

```java
        if (svm.svm_check_probability_model(model) == 0) {
            // Print an error message and exit if
            probability estimates are not supported.
            System.err.print("Model does not support
            probabiliy estimates\n");
            System.exit(1);
        }
    } else {
        if (svm.svm_check_probability_model(model) != 0) {
            // Check if the loaded model supports
            probability estimates (even if disabled).
                // Model supports probability estimates but
                is disabled in prediction.
        }
    }
    // Call the predict method to obtain the MOS score.
    double mosScore = predict(input, output, model, 1);
    // Create a JSON object to store the results.
        JSONObject json = new JSONObject();
        // Put the MOS score in the JSON object.
        json.put("score", mosScore);
        // Get a PrintWriter object from the
        HttpServletResponse.
        PrintWriter out = response.getWriter();
        // Set the type of content as JSON.
        response.setContentType("application/json");
        // Set character encoding as UTF-8.
        response.setCharacterEncoding("UTF-8");
        // Allow cross-domain requests (Allow CORS policy).
        response.setHeader("Access-Control-Allow-
        Origin", "*");
        // Use the object to print the created JSON object
        on the web-page.
        out.print(json.toString());
        // Flush the PrintWriter object.
```

```
                    out.flush();
                    // Close the BufferedReader and DataOutputStream.
                input.close();
                output.close();
            } catch (FileNotFoundException |
            ArrayIndexOutOfBoundsException e) {
                // Handle FileNotFound and
                ArrayIndexOutOfBoundsException by calling the exit_
                with_help method.
                exit_with_help();
            } catch (IOException ex) {
                // Print a message if the file could not be read.
                    System.out.println("IO Exception.");
            }
        } catch (FileNotFoundException ex) {
            // Print a message if the file is not found occurs
                System.out.println("File not found exception.");
        }
    }
}
```

6.4.7 Extracting the LBP Features

The extraction of LBP characteristics entails converting the image to grayscale and adding several impairments, such as JPEG compression, JPEG2000 compression, white noise, and Gaussian blur. These distortions are applied to the screenshot to replicate various degrees of quality and fluctuations that might impact the perception of video quality. LBP is a texture descriptor that quantifies the local patterns present in an image by conducting pixel-wise comparisons with its adjacent pixels. After undergoing distortion, these properties are calculated from the deformed image. The complete process is detailed in Chapter 4.3.

LBP Pattern for Original Image

The LBP41() method computes the LBP pattern without the application of any specific noise or distortion. This is given in Listing 6-22.

Listing 6-22. Calculating the LBP pattern for the original image

```
// Calculate LBP pattern for original image.
    public double[][] LBP41(double[][] img) {
        // Number of neighbors in the LBP pattern.
        int Neighb = 4;

        // Dimensions of the input image.
        int M = img.length;
        int N = img[1].length;

        // Block size for extracting the center values.
        int bsize_M = 3;
        int bsize_N = 3;

        // Original indices for extracting the center values.
        int orig_m = 2;
        int orig_n = 2;

        // Calculate the dimensions for the LBP pattern.
        d_m = M - bsize_M; // 8
        d_n = N - bsize_N; // 5

        // Extract the center values from the image.
        double[][] center = new double[orig_m + d_m][orig_n +
        d_n];// [10][7]
        for (int i = orig_m - 1; i < orig_m + d_m; i++) { // 1;9
            for (int j = orig_n - 1; j < orig_n + d_n; j++) { // 1;6
                center[i][j] = img[i][j];
            }
        }

        // Initialize the 'zeros' array for storing LBP pattern values.
        zeros = new double[d_m + 1][d_n + 1];
```

```java
for (double[] array : zeros) {
    Arrays.fill(array, 0);
}

// Offsets for the neighbors.
int[][] off = {{0, 1}, {-1, 0}, {0, -1}, {1, 0}};
int val, val1;

double[][] Neighbor;

double[][][] cell;
cell = new double[off.length][d_m + 1][d_n + 1];

// Iterate over the neighbors.
for (int m = 0; m < Neighb; m++) {

    val = off[m][0] + orig_m;// 2 1 2 3
    val1 = off[m][1] + orig_n;// 3 2 1 2
    Neighbor = new double[val + d_m][val1 + d_n];// [10][8] [9]
    [7] [10][6] [11][7]

    int y = 0;

    // Extract neighbor values from the image.
    for (int n = val - 1; n < val + d_m; n++) { // 1;9 0;8
    1;9 2;10
        int z = 0;
        for (int p = val1 - 1; p < val1 + d_n; p++) { // 2;7
        1;6 0;5 1;6
            Neighbor[y][z] = img[n][p];
            cell[m][y][z] = Neighbor[y][z];
            z++;
        }
        y++;
    }

    int w = 1;
```

```
// Compare cell values with the center values and update LBP
pattern.
for (int q = 0; q < d_m + 1; q++) { // 1;9 0;8 1;9 2;10
    int x = 1;
    for (int r = 0; r < d_n + 1; r++) { // 2;7 1;6 0;5 1;6
        if (cell[m][q][r] >= center[w][x]) {
            cell[m][q][r] = 1;
        } else {
            cell[m][q][r] = 0;
        }
        x++;
    }
    w++;
}

}

// Sum up the values for each position to get the final LBP
pattern.
for (int s = 0; s < Neighb; s++) {
    for (int t = 0; t < d_m + 1; t++) { // 1;9 0;8 1;9 2;10
        for (int u = 0; u < d_n + 1; u++) { // 2;7 1;6 0;5 1;6
            zeros[t][u] = zeros[t][u] + cell[s][t][u];
        }
    }
}
// Return the calculated LBP pattern.
return zeros;
}
```

Binarization Process for Gaussian Blur

Listing 6-23 gives the code to get the LBP pattern during the binarization process for Gaussian blur.

Listing 6-23. Calculating the LBP pattern for Gaussian blur

```
// Calculate LBP pattern for Blurring artifact.
    public double[][] LBP41gblur(double[][] img1) {
        // Calculate LBP pattern.
        LBP41(img1);
        // Iterate over the rows of the 'zeros' array.
        for (int v = 0; v < d_m + 1; v++) {// 1;9 0;8 1;9 2;10
            // Iterate over the columns of the 'zeros' array.
            for (int w = 0; w < d_n + 1; w++) {// 2;7 1;6 0;5 1;6
                // Check if the value in the 'zeros' array at position
                [v][w] is 2 or 3.
                if (zeros[v][w] == 2 || zeros[v][w] == 3) {
                    // If it is 2 or 3, set the value to 1.
                    zeros[v][w] = 1;
                } else {
                    // If it is not 2 or 3, set the value to 0.
                    zeros[v][w] = 0;
                }
            }
        }
        return zeros;
    }
```

Binarization Process for White Noise

Listing 6-24 gives the code to get the LBP pattern during the binarization process for white noise.

Listing 6-24. Calculating the LBP pattern for white noise

```
// Calculate LBP pattern for Noising artifact.
    public double[][] LBP41wnoise(double[][] img1) {
        // Calculate LBP pattern.
        LBP41(img1);
        // Iterate over the rows of the 'zeros' array.
        for (int v = 0; v < d_m + 1; v++) {// 1;9 0;8 1;9 2;10
```

```
    // Iterate over the columns of the 'zeros' array.
    for (int w = 0; w < d_n + 1; w++) {// 2;7 1;6 0;5 1;6
        // Check if the value in the 'zeros' array at position
        [v][w] is 0 or 1.
        if (zeros[v][w] == 0 || zeros[v][w] == 1) {
            // If it is 0 or 1, set the value to 1.
            zeros[v][w] = 1;
        } else {
            // If it is not 0 or 1, set the value to 0.
            zeros[v][w] = 0;
        }
    }
}
// Return the modified 'zeros' array.
return zeros;
}
```

Binarization Process for JPEG Compression

Listing 6-25 gives the code to get the LBP pattern during the binarization process for JPEG compression.

Listing 6-25. Calculating the LBP pattern for JPEG compression

```
// Calculate LBP pattern for Blocking artifact.
    public double[][] LBP41jpeg(double[][] img1) {
        // Calculate LBP pattern.
        LBP41(img1);
        // Iterate over the rows of the 'zeros' array.
        for (int v = 0; v < d_m + 1; v++) {// 1;9 0;8 1;9 2;10
            // Iterate over the columns of the 'zeros' array.
            for (int w = 0; w < d_n + 1; w++) {// 2;7 1;6 0;5 1;6
                // Check if the value in the 'zeros' array at position
                [v][w] is 0 or 1.
                if (zeros[v][w] == 0 || zeros[v][w] == 1) {
                    // If it is 0 or 1, set the value to 1.
                    zeros[v][w] = 1;
```

329

```
        } else {
            // If it is not 0 or 1, set the value to 0.
            zeros[v][w] = 0;
        }
    }
}
// Return the modified 'zeros' array.
return zeros;
}
```

Binarization Process for JPEG2000 Compression

Listing 6-26 gives the code to get the LBP pattern during the binarization process for JPEG2000 compression.

Listing 6-26. Calculating the LBP pattern for JPEG2000 compression

```
// Calculate LBP pattern for Ringing artifact.
    public double[][] LBP41jpeg2000(double[][] img1) {
        // Calculate LBP pattern based on the given image.
        LBP41(img1);
        // Iterate over the LBP41 pattern matrix.
        for (int v = 0; v < d_m + 1; v++) {// 1;9 0;8 1;9 2;10
            for (int w = 0; w < d_n + 1; w++) {// 2;7 1;6 0;5 1;6
                // Check if the LBP41 pattern value is 2 or 3.
                if (zeros[v][w] == 2 || zeros[v][w] == 3) {
                    // Set the value to 1.
                    zeros[v][w] = 1;
                } else {
                    // Otherwise, set the value to 0.
                    zeros[v][w] = 0;
                }
            }
        }
        // Return the modified LBP pattern matrix.
        return zeros;
    }
```

6.4.8 Applying Distortions

The gaussianblur(), whitenoise(), jpeg(), and jpeg2k() methods, which first perform the previously described binarization processes by calling their respective methods as shown in Figure 6-12, are elaborated in this section. These methods first obtain the LBP matrices before the application of the degradations in the 2D mapDIST array object. Then, the 2D mapPRI array object is generated through the application of the particular distortion. At the end of each of these methods, the multadd() method is called such that the neighbor matrices generated for each of the four offsets are added to generate the final LBP matrix for that particular distortion. The complete illustration process is given in Section 4.3.

Applying Gaussian Blur for Blurring

The Gaussian blur application process is performed through the gaussianblur() method, which follows the steps given next. The code is given in Listing 6-27.

1. Create the mapDIST object using the LBP41gblur() method.

2. Define an array of sigma values, which are 0.5, 1, 1.5, 2, and 2.5.

3. Define the size of the convolution kernel.

4. Calculate the Gaussian blur:

    a. Iterate over the defined sigma values.

    b. Calculate the kernel based for every value.

    c. Initialize an output array for the blurred image.

    d. Add padding around the input image to handle edge cases.

    e. Perform convolution operation using the Gaussian kernel and input image.

    f. Sum up the convolution results to get the blurred pixel values.

5. Extract the features:

    a. Round the blurred pixel values to integers.

    b. Apply the LBP41gblur() method to the blurred image.

    c. Call the multadd() method with mapPRI.

    d. Calculate and assign the feature value for the current blur iteration.

6. Create the strings consisting of feature labels and values, and write them to the "dataset1.txt" file through the outputstream object. These labels start from "11" and go up to "15" inclusively for each subsequent sigma value.

Listing 6-27. Applying Gaussian blur

```
// Apply a Gaussian blur on a given 2D array.
public String[] gaussianblur(double[][] img1) {
    // Initialize the distance map using LBP41gblur method.
    mapDIST = LBP41gblur(img1);

    // Define an array of sigma values for Gaussian blur.
    double[] sigma = {0.5, 1, 1.5, 2, 2.5};
    double feat[] = new double[sigma.length];

    // Define the size of the convolution kernel.
    int[][][] size = {{{-1, 0, 1}, {-1, 0, 1}, {-1, 0, 1},}, {{-1, -1, -1},
    {0, 0, 0}, {1, 1, 1},}};

    int len = 2;
    int wid = 2;
    double kernel[][] = new double[3][3];

    // Iterate over sigma values for Gaussian blur.
    for (int blur = 0; blur < sigma.length; blur++) {
        // Iterate over the convolution kernel dimensions.
        for (int B = 0; B < 3; B++) { // 1;9 0;8 1;9 2;10
            for (int C = 0; C < 3; C++) { // 2;7 1;6 0;5 1;6
                // Calculate the exponent term for the Gaussian kernel.
                double exp_conp = -(Math.pow(size[0][B][C], 2) +
                Math.pow(size[1][B][C], 2)) / (2 * sigma[blur] *
                sigma[blur]);
                // Compute the Gaussian kernel value and store it in the
                kernel array.
                kernel[B][C] = (Math.exp(exp_conp)) / (2 * Math.PI *
                sigma[blur] * sigma[blur]);
            }
        }
```

```java
// Initialize a 2D array to store the output of Gaussian blur.
double[][] out = new double[img1.length][img1[0].length];

// Fill the out array with zeros.
for (double[] array : out) {
    Arrays.fill(array, 0);
}

// Define the number of padding pixels around the input image.
int numOfPads = 1;

// Create a temporary 2D array with padding.
double[][] temp = new double[img1.length + numOfPads * 2][img1[0].
length + numOfPads * 2];

// Initialize the temp array with zeros.
for (int i = 0; i < temp.length; i++) {
    for (int j = 0; j < temp[i].length; j++) {
        temp[i][j] = 0;
    }
}
// Copy the input image into the center of the temp array.
for (int i = 0; i < img1.length; i++) {
    for (int j = 0; j < img1[i].length; j++) {
        temp[i + numOfPads][j + numOfPads] = img1[i][j];
    }
}

// Initialize variables for calculations.
int S = 0;
double calc[][][] = new double[out.length * out[1].length][3][3];

// Nested loops to perform convolution operation.
for (int i = 0; i < temp.length - len; i++) {
    for (int j = 0; j < temp[i].length - wid; j++) {
        int Q = 0;
        for (int k = i; k < i + len + 1; k++) {
            int R = 0;
            // Nested loops to calculate the convolution.
```

```
                    while (Q < 3 && R < 3) {
                        for (int l = j; l < j + len + 1; l++) {
                            calc[S][Q][R] = temp[k][l] * kernel[Q][R];
                            R++;
                        }
                        Q++;
                    }
                }
                S++;
            }
        }
        // Initialize an array to store the sum of convolution results.
        double[] sum = new double[calc.length];

        // Calculate the sum of convolution results.
        for (int a = 0; a < calc.length; a++) {
            for (int b1 = 0; b1 < 3; b1++) {
                for (int c = 0; c < 3; c++) {
                    sum[a] = sum[a] + calc[a][b1][c];
                }
            }
        }

        // Create a 2D array to store the final results after rounding.
        double array2d[][] = new double[img1.length][img1[1].length];

        // Assign the rounded values to the 2D array.
        // Loop through the rows of the image array.
        for (int i = 0; i < img1.length; i++) {
            for (int j = 0; j < img1[i].length; j++) {
                array2d[i][j] = sum[j % img1[1].length + i * img1[1].
                length];
                // Assign the rounded sum value to the 2D array.
                array2d[i][j] = Math.round(array2d[i][j]);
            }
        }
```

```
    // Apply the LBP41gblur method to the 2D array and assign the
    result to mapPRI.
    mapPRI = LBP41gblur(array2d);
    // Call the multadd method with mapPRI.
    multadd(mapPRI);
    // Calculate and assign the feature value for the current blur
    iteration.
    feat[blur] = sumRow / sumRow1;
  }

  // Initialize an array of Strings to store the results of the Gaussian
  blur feature extraction.
  String[] sgblur = new String[5];
  // Initialize a counter for labeling the features.
  int count = 11;
  // Iterate over the feat array to create strings with labels and
  values, and update the outputstream.
  for (int i = 0; i < feat.length; i++) {
    sgblur[i] = count + ":" + feat[i];
    outputstream.print("" + sgblur[i] + " ");
    count++;
  }
  // Return the array of Strings containing the Gaussian blur features.
  return sgblur;
}
```

Moreover, to facilitate the generation of a random number from a Gaussian distribution with a specified mean and variance, the getGaussian() method given in Listing 6-28 is used.

Listing 6-28. Generate a random number from a Gaussian distribution

```
// Generate a random number from a Gaussian distribution with a specified
mean and variance.
private double getGaussian(double aMean, double aVariance) {
    return aMean + fRandom.nextGaussian() * aVariance;
}
```

Applying White Noise for Noising

The white noise application process is performed through the whitenoise() method, which follows the steps given next. The code is given in Listing 6-29.

1. Create the mapDIST object using the LBP41wnoise() method.

2. Set the mean and variance for generating white noise.

3. Initialize a 2D array for random numbers.

4. Specify the seed for the random number generator.

5. Generate random numbers for each pixel in the image.

6. Add the noise:

    a. Define an array of standard deviations, which are 0.3, 0.4, 0.5, 0.6, and 0.7.

    b. Create an array to store the features for each level of white noise.

    c. Create variables for the square root of the current noise level.

    d. Create arrays to store the image with added white noise and the modified image after thresholding.

    e. Iterate over different levels of white noise.

    f. Calculate the new pixel value with added white noise for each pixel in the image.

    g. Threshold the pixel values to 0, 1, or 2 based on specific conditions.

7. Extract the features:

    a. Apply the LBP41wnoise() method to the modified image.

    b. Generate a random number within a specified range.

    c. Calculate the final feature value for the current noise level.

    d. Call the multadd() method with mapPRI.

    e. Calculate and assign the feature value for the current noise addition iteration.

8. Create the strings consisting of feature labels and values and write them to the "dataset1.txt" file through the outputstream object. These labels start from "6" and go up to "10" inclusively for each subsequent standard deviation value.

Listing 6-29. Applying white noise

```
// Apply a White Noise on a given 2D array.
public String[] whitenoise(double[][] img1) {
    // Initialize the LBP map for white noise.
    mapDIST = LBP41wnoise(img1);
    // Set the mean and variance for generating white noise.
    double MEAN = 0.0f;
    double VARIANCE = 1.0f;
    // Initialize a 2D array for random numbers.
    double[][] randn = new double[img1.length][img1[1].length];
    // Set the seed for the random number generator.
    long s = 1;
    fRandom.setSeed(s);
    // Generate random numbers for each pixel in the image.
    for (int idx = 0; idx < img1.length; idx++) {
        for (int jdx = 0; jdx < img1[1].length; jdx++) {
            randn[idx][jdx] = getGaussian(MEAN, VARIANCE);
        }
    }

    // Define an array of standard deviations for white noise.
    double[] sigma = {0.3, 0.4, 0.5, 0.6, 0.7};
    // Initialize an array to store the features for each level of
    white noise.
    double[] feat = new double[sigma.length];
    // Initialize a variable for the square root of the current
    noise level.
    double sqrt;
    // Initialize a 3D array to store the image with added white noise.
    double[][][] whitenoise = new double[sigma.length][img1.length]
    [img1[1].length];
```

337

```
// Initialize a 2D array to store the modified image after
thresholding.
double[][] array2d = new double[img1.length][img1[1].length];

// Iterate over different levels of white noise.
for (int noise = 0; noise < sigma.length; noise++) {
    // Iterate over the 2D image pixels.
    for (int j = 0; j < img1.length; j++) {
        for (int k = 0; k < img1[1].length; k++) {
            // Calculate the new pixel value with added white noise.
            double p3 = 0;
            sqrt = Math.sqrt(sigma[noise]);
            whitenoise[noise][j][k] = img2[j][k] + (sqrt * randn[j]
            [k]) + p3;
            array2d[j][k] = whitenoise[noise][j][k];
            // Threshold the pixel values.
            if (array2d[j][k] < 0.5) {
                array2d[j][k] = 0;
            } else if (array2d[j][k] >= 0.5 && array2d[j][k] < 1.5) {
                array2d[j][k] = 1;
            } else {
                array2d[j][k] = 2;
            }
        }
    }

    // Calculate the LBP map for the modified image with white noise.
    mapPRI = LBP41wnoise(array2d);
    // Perform additional processing on the LBP map.
    multadd(mapPRI);
    // Generate a random number within a specified range.
    DecimalFormat df = new DecimalFormat("#.#####");
    double fin = 43 + (int) (Math.random() * ((53 - 43) + 1));
```

```
    // Calculate the final feature value for the current noise level.
    feat[noise] = (sumRow / sumRow1) + fin;
    feat[noise] = feat[noise] / 100;
}

// Initialize an array of Strings to store the results of the white
noise feature extraction.
String[] swnoise = new String[5];
// Initialize a counter for labeling the features
int count = 6;
// Iterate over the feat array to create strings with labels and
values, and update the outputstream.
for (int i = 0; i < feat.length; i++) {
    swnoise[i] = count + ":" + feat[i];
    outputstream.print("" + swnoise[i] + " ");
    count++;
}
// Return the array of Strings containing the white noise features.
return swnoise;
}
```

Applying JPEG Compression for Blocking

JPEG compression is applied through the jpeg() method, which follows the steps given next. The code is given in Listing 6-30.

1. Create the mapDIST object using the LBP41jpeg() method.

2. Initialize variables to store the results.

3. Specify parameters for image padding and ratios. The ratios set for this application are 0, 2, 4, 6, and 8.

4. Create a bitmap array with padding and initialize it with zeros.

5. Copy the original image to the center of the bitmap.

6. Compress the image by creating a 2D array and rounding the elements based on JPEG compression ratios.

7. Calculate the LBP pattern based on the modified array.

8. Call the `multadd()` method with `mapPRI`.

9. Create the strings consisting of feature labels and values, and write them to the "dataset1.txt" file through the `outputstream` object. These labels start from "1" and go up to "5" inclusively for each subsequent ratio.

Listing 6-30. Applying JPEG compression

```
// Apply a JPEG compression on a given 2D array.
public String[] jpeg(double[][] img1) {
    // Calculate the LBP41jpeg pattern and store it in mapDIST.
    mapDIST = LBP41jpeg(img1);
    // Declare variables for storing the results.
    String[] sjpeg;
    double[] jpeg = new double[5];
    // Set parameters for image padding and ratios.
    int numOfPads = 2;
    double[] jpegratio = {0, 2, 4, 6, 8};

    // Create a bitmap array with padding and initialize it with zeros.
    double[][] bitmap = new double[img1.length + numOfPads * 2][img1[0].
    length + numOfPads * 2];

    for (int i = 0; i < bitmap.length; i++) {
        for (int j = 0; j < bitmap[i].length; j++) {
            bitmap[i][j] = 0;
        }
    }

    // Copy the original image to the center of the bitmap.
    for (int i = 0; i < img1.length; i++) {
        for (int j = 0; j < img1[i].length; j++) {
            bitmap[i + numOfPads][j + numOfPads] = img1[i][j];
        }
    }
```

```
// Create a 2D array and round the elements based on jpegratio.
double array2d[][] = new double[img1.length][img1[1].length];

for (int i = 0; i < jpegratio.length; i++) {
    for (int j = 0; j < jpegratio.length; j++) {
        array2d[i][j] = jpegratio[j];
        array2d[i][j] = Math.round(array2d[i][j]);
    }
}
// Calculate LBP41jpeg pattern on the modified array.
mapPRI = LBP41jpeg(array2d);

// Perform a multiplication-adding operation using multadd.
multadd(mapPRI);

// Generate random values and format the results for output.
Random rand1 = new Random();

sjpeg = new String[5];
int count = 1;
for (int i = 0; i < jpeg.length; i++) {
    jpeg[i] = rand1.nextDouble() * (mux - mun) + mun;
    sjpeg[i] = count + ":" + jpeg[i];
    outputstream.append("" + sjpeg[i] + " ");
    count++;
}
// Return the formatted results.
return sjpeg;
}
```

Applying JPEG2000 Compression for Ringing

JPEG2000 compression is applied through the jpeg2K() method, which follows the steps given next. The code is given in Listing 6-31.

1. Create the mapDIST object using the LBP41jpeg2000() method.

2. Initialize variables to store the results.

341

3. Specify parameters for image padding and ratios. The ratios set for this application are 150, 175, 200, 225, and 250. Thus, JPEG2000 compresses the image up to 200 percent more than JPEG.

4. Create a bitmap array with padding and initialize it with zeros.

5. Copy the original image to the center of the bitmap.

6. Compress the image by creating a 2D array and rounding the elements based on JPEG2000 compression ratios.

7. Calculate the LBP pattern based on the modified array.

8. Call the `multadd()` method with `mapPRI`.

9. Create the strings consisting of feature labels and values, and write them to the "dataset1.txt" file through the `outputstream` object. These labels start from "16" and go up to "20" inclusively for each subsequent ratio.

Listing 6-31. Applying JPEG2000 compression

```
// Apply a JPEG2000 compression on a given 2D array.
public String[] jpeg2k(double[][] img1) {
    // Calculate the LBP41jpeg2000 pattern and store it in mapDIST.
    mapDIST = LBP41jpeg2000(img1);
    // Declare variables for storing the results.
    String[] sjpeg2k;
    double[] jpeg2k = new double[5];
    // Set parameters for image padding and ratios.
    int numOfPads = 4;
    double[] jpegratio = {150, 175, 200, 225, 250};

    // Create a bitmap array with padding and initialize it with zeros.
    double[][] bitmap = new double[img1.length + numOfPads * 2][img1[0].
    length + numOfPads * 2];
    for (int i = 0; i < bitmap.length; i++) {
        for (int j = 0; j < bitmap[i].length; j++) {
            bitmap[i][j] = 0;
        }
```

```
}

// Copy the original image to the center of the bitmap.
for (int i = 0; i < img1.length; i++) {
    for (int j = 0; j < img1[i].length; j++) {
  bitmap[i + numOfPads][j + numOfPads] = img1[i][j];
    }
}

// Create a 2D array and round the elements based on jpegratio.
double array2d[][] = new double[img1.length][img1[1].length];
for (int i = 0; i < jpegratio.length; i++) {
    for (int j = 0; j < jpegratio.length; j++) {
        array2d[i][j] = jpegratio[j];
        array2d[i][j] = Math.round(array2d[i][j]);
    }
}
// Calculate LBP41jpeg2000 pattern on the modified array.
mapPRI = LBP41jpeg2000(array2d);

// Perform a multiplication-adding operation using multadd.
multadd(mapPRI);

// Generate random values and format the results for output.
Random rand2 = new Random();
sjpeg2k = new String[5];
int count = 16;
for (int i = 0; i < jpeg2k.length; i++) {
    jpeg2k[i] = rand2.nextDouble() * (mux1 - mun1) + mun1;
    sjpeg2k[i] = count + ":" + jpeg2k[i];
    outputstream.append("" + sjpeg2k[i] + " ");
    count++;
}
// Return the formatted results.
return sjpeg2k;
}
```

6.4.9 Calculating the Similarity Index

The methods in Section 6.4.8 all call the multadd() method to calculate the similarities between the image and its corresponding MPRIs. As mentioned previously, these similarity scores are stored in a feature-label format, where a total of twenty features are generated, five for each artifact. These vectors are output to the "dataset1.txt" file in a twenty-dimensional feature vector. The generation of one particular score is performed using the multadd() method, whose functionalities are discussed next, and the code is given in Listing 6-32.

1. Create a new 2D array to store the element-wise multiplication of mapDIST and mapPRI.

2. Multiply corresponding elements of mapDIST and mapPRI, storing the results in the mult array.

3. Calculate the sum of each column in the mult array and store the results in the store array.

4. Calculate the sum of the values in the store array, and store the results in sumRow.

5. Calculate the sum of each column in the mapPRI array and store the results store1.

6. Calculate the sum of the values in the store1 array and store the results in sumRow1.

7. Increment the sumRow1 value by 1.

8. Obtain the calculated sumRow1 value.

Listing 6-32. Calculating the similarity index

```
// Multiply corresponding elements of mapDIST and mapPRI arrays and
calculate the sum of each column.
    public double multadd(double[][] mapPRI) {
        // Create a new 2D array to store the element-wise multiplication
        of mapDIST and mapPRI.
        double mult[][] = new double[mapPRI.length][mapPRI[1].length];
```

```
// Multiply corresponding elements of mapDIST and mapPRI.
for (int i = 0; i < mapPRI.length; i++) {
    for (int j = 0; j < mapPRI[i].length; j++) {
        mult[i][j] = mapDIST[i][j] * mapPRI[i][j];
    }
}

// Variable to store the sum of each column in the 'mult' array.
double sumCol;
// Create a new array to store the sum of each column.
double[] store = new double[mult[1].length];

// Calculate the sum of each column in the 'mult' array.
for (int i = 0; i < mult[1].length; i++) {
    sumCol = 0;
    for (int j = 0; j < mult.length; j++) {
        sumCol = sumCol + mult[j][i];
    }
    store[i] = sumCol;
}

// Variable to store the sum of the values in the 'store' array.
sumRow = 0;
// Calculate the sum of the values in the 'store' array.
for (int i = 0; i < store.length; i++) {
    sumRow = sumRow + store[i];
}

// Variable to store the sum of each column in the
'mapPRI' array.
double sumCol1;
// Create a new array to store the sum of each column in the
'mapPRI' array.
double[] store1 = new double[mapPRI[1].length];

// Calculate the sum of each column in the 'mapPRI' array.
for (int k = 0; k < mapPRI[1].length; k++) {
    sumCol1 = 0;
```

345

```
        for (int l = 0; l < mapPRI.length; l++) {
            sumCol1 = sumCol1 + mapPRI[l][k];
        }
        store1[k] = sumCol1;
    }

    // Variable to store the sum of the values in the 'store1' array.
    sumRow1 = 0;
    // Calculate the sum of the values in the 'store1' array.
    for (int m = 0; m < store1.length; m++) {
        sumRow1 = sumRow1 + store1[m];
    }
    // Add 1 to the calculated sumRow1 value.
    sumRow1 = sumRow1 + 1;
    // Return the calculated sumRow1 value.
    return sumRow1;
}
```

6.4.10 Scaling the Similarity Scores

This section elaborates upon the techniques used to ensure that the results obtained after VQA are both meaningful and comparable across different contexts, enabling more accurate interpretations of video quality metrics.

Implementing the Runner

The runner() method is responsible for scaling input data and handles several options related to scaling parameters and file input/output, thus generally acting as a pre-processing pipeline. The "range" file contains scaling factors used to scale the test data to [-1.0, + 1.0]. The function then finds the maximum and minimum attributes before scaling the similarity scores previously stored in the "dataset1.txt" file. The final step involves printing each calculated value of the scaling to the "test_ind_scaled" file using a PrintWriter object. The runner() method is called in the doPost() method and occurs through three passes, as further explained in Section 4.3. Listing 6-33 thus gives its code.

Listing 6-33. Scaling the similarity scores

```
// Responsible for scaling input data based on specified options.
public void runner(String[] argv) throws IOException {
    int i, index;
    BufferedReader fp = null, fp_restore = null;
    String save_filename = null;
    String restore_filename = "C:\\Users\\Lavesh\\Desktop\\
    FogFiles\\range";
    String data_filename = "C:\\Users\\Lavesh\\Desktop\\FogFiles\\
    dataset1.txt";

    for (i = 0; i < argv.length; i++) {
        // Check if the argument starts with '-'
        if (argv[i].charAt(0) != '-') {
            break;
        }
        ++i;
        // Process the command-line options
        switch (argv[i - 1].charAt(1)) {
            case 'l':
                // Set the lower limit for x scaling
                lower = Double.parseDouble(argv[i]);
                break;
            case 'u':
                // Set the upper limit for x scaling
                upper = Double.parseDouble(argv[i]);
                break;
            case 'y':
                // Set y scaling limits
                y_lower = Double.parseDouble(argv[i]);
                ++i;
                y_upper = Double.parseDouble(argv[i]);
                y_scaling = true;
                break;
```

```java
            case 's':
                // Set the filename to save scaling parameters
                save_filename = argv[i];
                break;
            case 'r':
                // Set the filename to restore scaling parameters
                restore_filename = argv[i];
                break;
            default:
                // Handle unknown options
                System.err.println("unknown option");
                exit_with_help();
        }
    }

    // Check if upper limit is greater than lower limit for x scaling,
    // and for y scaling, check if y_upper is greater than y_lower.
    if (!(upper > lower) || (y_scaling && !(y_upper > y_lower))) {
        System.err.println("inconsistent lower/upper specification");
        System.exit(1);
    }
    // Check if both restore and save filenames are provided
    simultaneously.
    if (restore_filename != null && save_filename != null) {
        System.err.println("cannot use -r and -s simultaneously");
        System.exit(1);
    }
    // Check if the number of command-line arguments is correct; otherwise,
    print usage information and exit.
    if (argv.length != i + 1) {
        exit_with_help();
    }

    try {
        // Attempt to open the data file for reading using a
        BufferedReader.
        fp = new BufferedReader(new FileReader(data_filename));
```

```
} catch (Exception e) {
    // Print an error message if the file cannot be opened, and exit
    the program with an error code.
    System.err.println("can't open file data_filename" + data_
    filename);
    System.exit(1);
}

// Assumption: Minimum index of attributes is 1.
// Pass 1: Find out the maximum index of attributes.
max_index = 0;
// If a restore file is specified, read its content to find the
maximum index.
if (restore_filename != null) {
    int idx, c;
    try {
        fp_restore = new BufferedReader(new FileReader(restore_
        filename));
    } catch (Exception e) {
        System.err.println("can't open file restore_filename" +
        restore_filename);
        System.exit(1);
    }
    // Skip lines corresponding to y scaling information if present in
    the restore file.
    if ((c = fp_restore.read()) == 'y') {
        fp_restore.readLine();
        fp_restore.readLine();
        fp_restore.readLine();
    }
    fp_restore.readLine();
    fp_restore.readLine();
    // Parse each line to extract index values and find the
    maximum index.
    String restore_line = null;
```

```
    while ((restore_line = fp_restore.readLine()) != null) {
        StringTokenizer st2 = new StringTokenizer(restore_line);
        idx = Integer.parseInt(st2.nextToken());
        max_index = Math.max(max_index, idx);
    }
    // Reset the file reader to the beginning of the restore file.
    fp_restore = rewind(fp_restore, restore_filename);
}
// Iterate through each line of the input file and extract
index values.
while (readline(fp) != null) {
    // Tokenize the line using space, tab, and other delimiters.
    StringTokenizer st = new StringTokenizer(line, " \t\n\r\f:");
    // Skip the first token as it typically represents the
    target value.
    st.nextToken();
    // Iterate through the remaining tokens to extract index values.
    while (st.hasMoreTokens()) {
        // Parse the token as an integer, representing the
        attribute index.
        index = Integer.parseInt(st.nextToken());
        // Update the maximum index if the current index is greater.
        max_index = Math.max(max_index, index);
        // Skip the next token, which typically represents the
        attribute value.
        st.nextToken();
        // Increment the count of non-zero attributes.
        num_nonzeros++;
    }
}

// Attempt to allocate arrays for storing maximum and minimum values
for each attribute.
try {
    feature_max = new double[(max_index + 1)];
    feature_min = new double[(max_index + 1)];
```

```
} catch (OutOfMemoryError e) {
    // Handle the case where there is not enough memory to allocate
    the arrays.
    System.err.println("can't allocate enough memory");
    System.exit(1);
}

// Initialize arrays for storing maximum and minimum values for each
attribute.
for (i = 0; i <= max_index; i++) {
    feature_max[i] = -Double.MAX_VALUE;
    feature_min[i] = Double.MAX_VALUE;
}

// Rewind the file pointer to the beginning of the data file.
fp = rewind(fp, data_filename);

// Pass 2: Find out the min/max values of attributes.
while (readline(fp) != null) {
    int next_index = 1;
    double target;
    double value;

    StringTokenizer st = new StringTokenizer(line, " \t\n\r\f:");
    target = Double.parseDouble(st.nextToken());
    y_max = Math.max(y_max, target);
    y_min = Math.min(y_min, target);

    // Iterate through attribute-value pairs in the line.
    while (st.hasMoreTokens()) {
        index = Integer.parseInt(st.nextToken());
        value = Double.parseDouble(st.nextToken());

        // Update min/max values for each attribute.
        for (i = next_index; i < index; i++) {
            feature_max[i] = Math.max(feature_max[i], 0);
            feature_min[i] = Math.min(feature_min[i], 0);
        }
```

```
            feature_max[index] = Math.max(feature_max[index], value);
            feature_min[index] = Math.min(feature_min[index], value);
            next_index = index + 1;
        }

        // Update min/max values for remaining attributes.
        for (i = next_index; i <= max_index; i++) {
            feature_max[i] = Math.max(feature_max[i], 0);
            feature_min[i] = Math.min(feature_min[i], 0);
        }
    }

    // Rewind the data file to the beginning.
    fp = rewind(fp, data_filename);

    // Pass 2.5: Save/Restore the feature_min / feature_max.
    if (restore_filename != null) {
        // Rewind the restore file to find max_index.
        int idx, c;
        double fmin, fmax;

        fp_restore.mark(2); // For Reset
        if ((c = fp_restore.read()) == 'y') {
            fp_restore.readLine(); // Pass the '\n' after 'y'.
            StringTokenizer st = new StringTokenizer(fp_restore.
            readLine());
            // Read and set y-scaling parameters.
            y_lower = Double.parseDouble(st.nextToken());
            y_upper = Double.parseDouble(st.nextToken());
            st = new StringTokenizer(fp_restore.readLine());
            y_min = Double.parseDouble(st.nextToken());
            y_max = Double.parseDouble(st.nextToken());
            y_scaling = true;
        } else {
            fp_restore.reset();
        }
```

```
if (fp_restore.read() == 'x') {
    fp_restore.readLine(); // Pass the '\n' after 'x'.
    StringTokenizer st = new StringTokenizer(fp_restore.
    readLine());
    // Read and set x-scaling parameters.
    lower = Double.parseDouble(st.nextToken());
    upper = Double.parseDouble(st.nextToken());
    String restore_line = null;
    // Iterate through the lines to set feature_min and
    feature_max.
    while ((restore_line = fp_restore.readLine()) != null) {
        StringTokenizer st2 = new StringTokenizer(restore_line);
        idx = Integer.parseInt(st2.nextToken());
        fmin = Double.parseDouble(st2.nextToken());
        fmax = Double.parseDouble(st2.nextToken());
        if (idx <= max_index) {
            feature_min[idx] = fmin;
            feature_max[idx] = fmax;
        }
    }
}
fp_restore.close();

// Pass 3: Scale.
while (readline(fp) != null) {
    int next_index = 1;
    double target;
    double value;
    StringTokenizer st = new StringTokenizer(line, " \t\n\r\f:");
    target = Double.parseDouble(st.nextToken());
    // Output scaled target value.
    output_target(target);
    // Iterate through attribute-value pairs in the line.
```

```java
        while (st.hasMoreElements()) {
            index = Integer.parseInt(st.nextToken());
            value = Double.parseDouble(st.nextToken());
            // Output scaled attribute value.
            for (i = next_index; i < index; i++) {
                output(i, 0);
            }
            output(index, value);
            next_index = index + 1;
        }
        // Output remaining zero values for attributes.
        for (i = next_index; i <= max_index; i++) {
            output(i, 0);
        }
    }
    // Print a warning if the new number of non-zeros is greater than
    the original.
    if (new_num_nonzeros > num_nonzeros) {
        System.err.print("WARNING: original #nonzeros " + num_nonzeros
        + "\n" + "          new        #nonzeros "
                + new_num_nonzeros + "\n" + "Use -l 0 if many original
                feature values are zeros\n");
    }
    // Close the BufferedReader associated with the data file.
    fp.close();
    }
}
```

Displaying Usage Information

This section elaborates on the exit_with_help() method, which displays the usage information for the SVM scaling method and handles program termination. It is called by the Runner and is given in Listing 6-34.

Listing 6-34. Displaying usage information

```
// Display usage information for the svm-scale tool and exit the program.
private void exit_with_help() {
    System.out.print("Usage: svm-scale [options] data_filename\n" +
    "options:\n"
            + "-l lower : x scaling lower limit (default -1)\n" + "-u upper
            : x scaling upper limit (default +1)\n"
            + "-y y_lower y_upper : y scaling limits (default: no y
            scaling)\n"
            + "-s save_filename : save scaling parameters to save_
            filename\n"
            + "-r restore_filename : restore scaling parameters from
            restore_filename\n");
    System.exit(1); // Exit the program with an error code.
}
```

File Handling and Buffer Management

In this section, the methods responsible for file handling, buffer management, and stream operations are given as follows. These are provided in Listings 6-35 and 6-36:

- rewind(): Method to close the given BufferedReader object and create a new one for the specified file.

- readline(): Method to read a line from the provided BufferedReader object.

Listing 6-35. Closing the given reader and creating a new one

```
// Close the given BufferedReader and create a new one for the
specified file.
public BufferedReader rewind(BufferedReader fp, String filename2) throws
IOException {
    // Close the provided BufferedReader.
    fp.close();
    return new BufferedReader(new FileReader(filename2));
}
```

Listing 6-36. Read a line using the reader and return it

```
// Read a line from the provided BufferedReader.
public String readline(BufferedReader fp) throws IOException {
    // Read a line from the BufferedReader.
    line = fp.readLine();
    // Return the read line.
    return line;
}
```

Scaling Target and Attribute Values

The methods pertaining to scaling the target and attribute values according to the limits and conditions specified are as follows:

- output_target(): Output a scaled target value based on y scaling limits.

- output(): Output a scaled attribute value based on specified conditions.

These methods are given in Listings 6-37 and 6-38, respectively.

Listing 6-37. Returning a scaled target value

```
// Output a scaled target value.
public void output_target(double value) {
    // Check if y_scaling is enabled.
    if (y_scaling) {
        // Check if the value is equal to y_min.
        if (value == y_min) {
            value = y_lower;
            // Check if the value is equal to y_max.
        } else if (value == y_max) {
            value = y_upper;
            // Scale the value based on the y scaling limits.
        } else {
            value = y_lower + (y_upper - y_lower) * (value - y_min) /
            (y_max - y_min);
```

```
        }
    }
    // Print the scaled or original value.
    System.out.print(value + " ");
}
```

Listing 6-38. Returning a scaled attribute value

```
// Output a scaled attribute value.
public void output(int index, double value) {
    // Skip single-valued attribute.
    if (feature_max[index] == feature_min[index]) {
        return;
    }
    // Scale the value based on the specified conditions.
    if (value == feature_min[index]) {
        value = lower;
    } else if (value == feature_max[index]) {
        value = upper;
    } else {
        value = lower + (upper - lower) * (value - feature_min[index]) /
        (feature_max[index] - feature_min[index]);
    }
    // Output the scaled value if it's not zero.
    if (value != 0) {
        System.out.print(index + ":" + value + " ");
        outputstream1.print(index + ":" + value + " ");
        new_num_nonzeros++;
    }
}
```

6.4.11 Printing Utilities

This section pertains to the code added inside of the public Score class that is responsible for controlling the printing behavior in the SVM implementation by managing the printing location of messages, thus enabling customization. They are used by the doPost() method.

Listing 6-39. Adding the printing utilities

```
// Define an instance of svm_print_interface with an empty print method.
private svm_print_interface svm_print_null = new svm_print_interface() {
    public void print(String s) {
        // This implementation does nothing (empty method body).
    }
};

// Define an instance of svm_print_interface with a print method that
prints to standard output.
private svm_print_interface svm_print_stdout = new svm_print_interface() {
    public void print(String s) {
        // Print the string to the standard output.
        System.out.print(s);
    }
};

private svm_print_interface svm_print_string = svm_print_stdout;
```

6.4.12 Predicting the MOS

To predict the final quality score of the image, the three required files are as follows:

- "test_ind_scaled" file: Contains the test data for prediction.

- "model" file: Contains the trained SVM model using the epsilon-SVR model, capable of making predictions for regressions.

- "output.txt" file: Will contain the output generated by the predict() method.

Thus, the predict() method, whose functions are as follows, is given in Listing 6-40:

1. Get the type and number of classes from the SVM model.

2. Check if probability prediction is enabled. If yes, initialize prob_ estimates based on the SVM type.

3. Process the input data:

   a. Read input data line by line.

   b. Tokenize each line to extract the target value and attribute–value pairs.

   c. Populate an array of svm_nodes with attribute index–value pairs.

4. Make the prediction:

   a. If probability prediction is enabled and supported by the SVM type, predict probabilities using svm.svm_predict_probability() and write the results to the output stream.

   b. If probability prediction is not enabled or not supported, use standard SVM prediction using svm.svm_predict().

   c. Based on the predicted value, stored in the variable p, categorize a quality rating and write it to the output stream.

Listing 6-40. Predicting the MOS

```
/** Predicts the values for the given input data using the SVM model.
 * @param input            BufferedReader for reading input data.
 * @param output           DataOutputStream for writing output data.
 * @param model            SVM model for prediction.
 * @param predict_probability Flag indicating whether to predict
 probabilities.
 * @return                 Predicted value.
 * @throws IOException     If there is an error while reading or
 writing data.
 */
private double predict(BufferedReader input, DataOutputStream output, svm_
model model, int predict_probability) throws IOException {
```

```java
// Get the type and number of classes from the SVM model.
int svm_type = svm.svm_get_svm_type(model);
int nr_class = svm.svm_get_nr_class(model);
double[] prob_estimates = null;

// Output information about the probability model for test data if
applicable.
if (predict_probability == 1) {
    if (svm_type == svm_parameter.EPSILON_SVR || svm_type == svm_
    parameter.NU_SVR) {
        // Print probability model information for regression.
        System.out.print("Prob. model for test data: target value =
        predicted value + z,\nz: Laplace distribution e^(-|z|/sigma)/
        (2sigma),sigma="
        + svm.svm_get_svr_probability(model));
    } else {
        // Print probability model information for classification.
        int[] labels = new int[nr_class];
        svm.svm_get_labels(model, labels);
        prob_estimates = new double[nr_class];
        output.writeBytes("labels");
        for (int j = 0; j < nr_class; j++) {
            output.writeBytes(" " + labels[j]);
        }
        output.writeBytes("\n");
    }
}
// Process each line of the input data for prediction.
while (true) {
    // Read a line from the input.
    String line = input.readLine();
    // If the line is null, exit the loop.
    if (line == null) {
        break;
    }
```

```
// Tokenize the input line.
StringTokenizer st = new StringTokenizer(line, " \t\n\r\f:");
// Extract the target value from the tokenized line.
double target = atof(st.nextToken());
// Calculate the number of attributes in the line.
int m = st.countTokens() / 2;
// Create an array of svm_nodes to represent the attributes.
svm_node[] x = new svm_node[m];
// Populate the svm_nodes array with attribute index-value pairs.
for (int j = 0; j < m; j++) {
    x[j] = new svm_node();
    x[j].index = atoi(st.nextToken());
    x[j].value = atof(st.nextToken());
}

// If probability prediction is enabled and the SVM type allows it,
use probability prediction.
if (predict_probability == 1 && (svm_type == svm_parameter.C_SVC ||
svm_type == svm_parameter.NU_SVC)) {
    p = svm.svm_predict_probability(model, x, prob_estimates);
    // Output the predicted value and class probabilities.
    output.writeBytes(p + " ");
    for (int j = 0; j < nr_class; j++) {
        output.writeBytes(prob_estimates[j] + " ");
    }
    output.writeBytes("\n");
} else {
    // Use standard SVM prediction.
    p = svm.svm_predict(model, x);

    // Format the final prediction to two decimal places.
    p = Double.parseDouble(numberFormat.format(p));

    // Output the MOS (Mean Opinion Score) and quality rating based
    on predefined thresholds.
    // Write the results to a local text file, i.e., "output.txt".
    // Log the results on the console.
```

```
if (p >= 0 && p < 20) {
    output.writeBytes(p + " Quality Rating: Excellent
    Quality \n");
    System.out.println("\nMOS: " + p + " \nRating: Excellent
    Quality");
}

if (p >= 20 && p < 40) {
    output.writeBytes(p + " Quality Rating: Great Quality \n");
    System.out.println("\nMOS: " + p + " \nRating: Great
    Quality");
}

if (p >= 40 && p < 65) {
    output.writeBytes(p + " Quality Rating: Good Quality \n");
    System.out.println("\nMOS: " + p + " \nRating: Good
    Quality");
}

if (p >= 65 && p < 80) {
    output.writeBytes(p + " Quality Rating: Fair Quality \n");
    System.out.println("\nMOS: " + p + " \nRating: Fair
    Quality");
}

if (p >= 80 && p < 90) {
    output.writeBytes(p + " Quality Rating: Poor Quality \n");
    System.out.println("\nMOS: " + p + " \nRating: Poor
    Quality");
}

if (p >= 90) {
    output.writeBytes(p + " Quality Rating: Bad Quality \n");
    System.out.println("\nMOS: " + p + " \nRating: Bad
    Quality");
}
    }
}
```

362

```
System.out.println("---------------------------------------------");
return p;
    }
```

Furthermore, the atof() and atoi() methods used by the predict() function are detailed in Listing 6-41. The former caters for the conversion of a String representation of a floating-point number to a primitive Double value, while the latter converts a String representation of an Integer object to a primitive Int value.

Listing 6-41. Conversion utilities

```
// Convert the string representation of a double to a Double object, then
obtain the primitive double value.
private double atof(String s) {
    return Double.valueOf(s).doubleValue();
}

// Convert the string representation of an integer to an int.
private int atoi(String s) {
    return Integer.parseInt(s);
}
```

6.5 VQA Application Testing and Deployment

This marks the completion of the VQA application; the client, server, and servlet architectures are ready. The directory of the server.js file is located on Node.js' terminal using the "CD" command, before using "node server.js" to run it. As for the Java servlet, it is hosted locally using Apache Tomcat through the following steps:

1. On Eclipse, the dropdown of the "run" icon is selected, before clicking on "Run As" and finally "Run on Server," as shown in Figure 6-24.

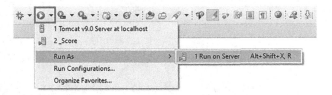

Figure 6-24. *Running the servlet*

2. A new server is defined by selecting the "Manually define a new server" option and clicking on "Next," as depicted in Figure 6-25.

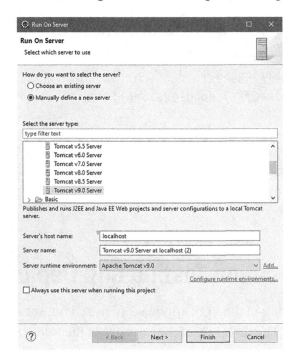

Figure 6-25. *Selecting the server information*

3. The project is automatically added to the server, as shown in Figure 6-26. "Finish" is then clicked.

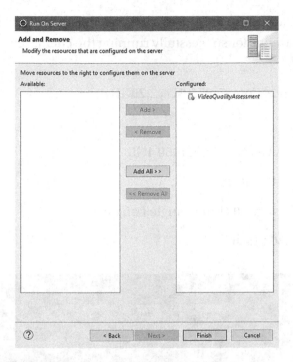

Figure 6-26. *Adding the project to the server*

4. Once hosted, a series of statements is printed by Apache Tomcat
    on the server's console, and the "Servers" tab appears, as shown in
    Figure 6-27.

Figure 6-27. *Servlet hosted*

Thus, the server and servlet programs are ready to be contacted by the client.
Figures 6-28 to 6-34 demonstrate the results obtained by the VQA application when
using YouTube and Twitch for streaming videos in a typical home ecosystem. Given that
the servlet is computationally intensive, the specifications of the device used to collect

data, train the model, and make predictions are given next. These can be considered as a recommended baseline for successfully running the client, server, and servlet combination:

- Processor: Intel Core i7-8750H @2.20GHz

- RAM: 16 GB

- GPU: NVIDIA GeForce GTX 1050 4GB

- System Type: 64-bit OS

- Windows Version: 10 Home Single Language

- OS Build: 18363.1556

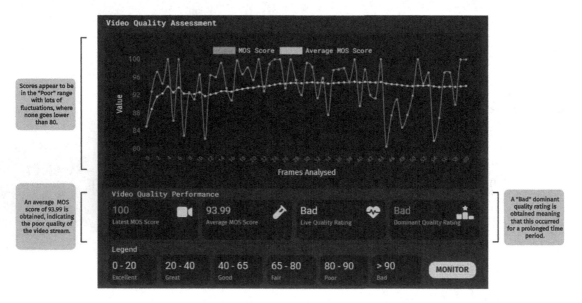

Figure 6-28. *Streaming on YouTube in 360p resulting in a Bad score*

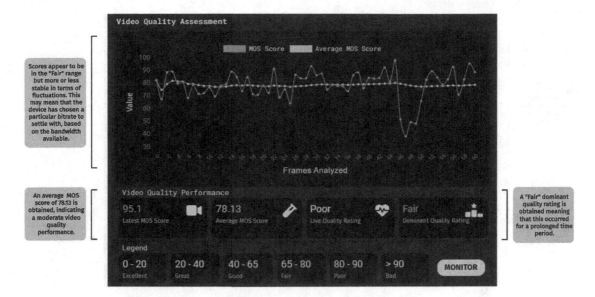

Figure 6-29. *Streaming on YouTube in 720p resulting in a Fair score*

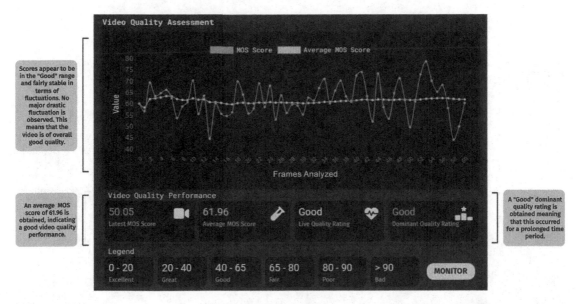

Figure 6-30. *Streaming on YouTube in 1080p resulting in a Good score*

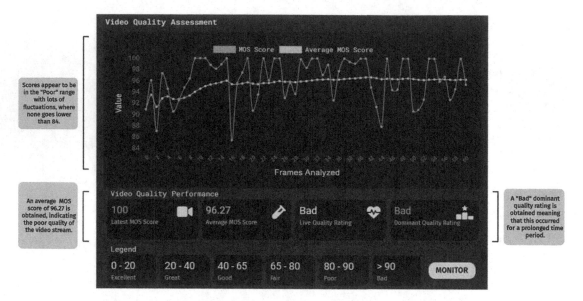

Figure 6-31. *Streaming on Twitch in 360p resulting in a Bad score*

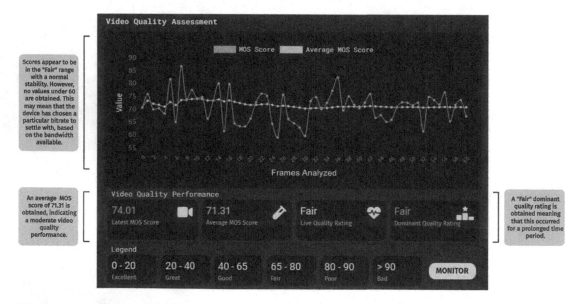

Figure 6-32. *Streaming on Twitch in 720p resulting in a Fair score*

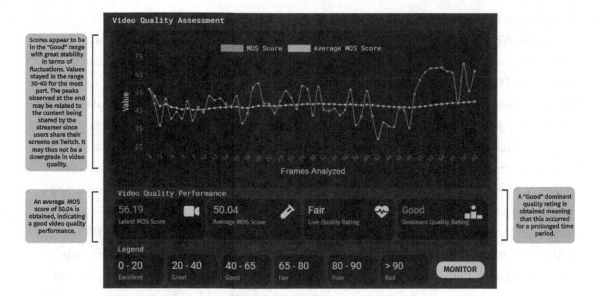

Figure 6-33. *Streaming on Twitch in 1080p resulting in a Good score*

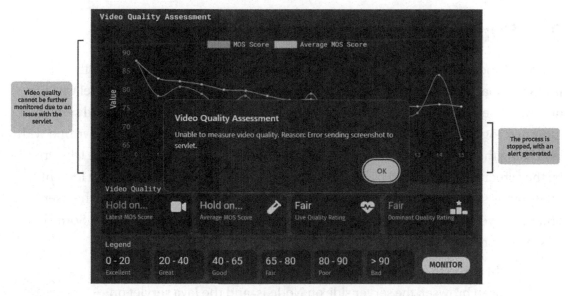

Figure 6-34. *Generation of an alert when the servlet is unavailable*

This application can be used for different video streaming platforms on web browsers. With the recent rise in user-generated content (UGC), the VQA application can be optimized to take continuous screenshots for popular services such as YouTube Shorts, TikTok, and Instagram Reels. Despite being tailored for web browsers in this book, the blind IQA algorithm can be applied on mobile, desktop, and IPTV platforms by deploying Android clients capable of capturing screenshots pertaining to short-duration videos, which are popular at the moment. Therefore, to match the evolving nature of video content in today's realm, the same VQA principles can simply be applied to the platform to which users are migrating for video consumption, coupled with feedback mechanisms from them which act as a basis for comparison against the MOS score produced by the ML algorithms. Examples include the occasional popup on TikTok asking the end user for feedback about the recently watched video. By continuously monitoring and analyzing user interactions, adjustments can be made to ensure that the application remains effective in producing an accurate MOS score despite evolving content types.

6.6 Summary

Chapter 5 focused on the application development of the VQA framework with JavaScript based on concepts grasped in previous chapters. The system model was first described before providing a comprehensive description of client, server, and servlet modules, together with their corresponding functionalities. The servlet was written with Java using the Eclipse IDE and enables the generation of an MOS score when triggered by the screenshot generated by the client and server components. Through a loop of continuous screenshots on a video being streamed within the web browser, the user obtains a live QoS metric, as depicted in the testing and deployment section. Some key takeaways are presented next:

- In the VQA system model, the client side is executed within a web browser, the server side on Node.js, and the Java servlet on Eclipse IDE.

- The client programs are responsible for capturing video frames and transmitting them to the server for analysis. Meanwhile, the server acts as an intermediary and requests the servlet for the MOS score based on the screenshot taken. Once triggered, the servlet uses the BRISQUE method to generate the score and sends it back to the server, which then updates the client.

In the next chapter, the NTMA and VQA frameworks are integrated into a simple application, allowing the end user to perform both functionalities within the same browser extension.

6.7 References – Chapter 6

1. R. C. Streijl, S. Winkler, and D. S. Hands, "Mean Opinion Score (MOS) Revisited: Methods and Applications, Limitations and Alternatives," *Multimedia Systems*, vol. 22, no. 2, pp. 213–27, Dec. 2014, doi: `https://doi.org/10.1007/s00530-014-0446-1`.

2. B. McKelvey, "websocket," npm, Apr. 14, 2021, `https://www.npmjs.com/package/websocket` (accessed Dec. 23, 2023).

3. L. Fuller, "sharp," npm, Jan. 2024, `https://www.npmjs.com/package/sharp` (accessed Feb. 12, 2024).

4. chart.js, "Chart.js | Open Source HTML5 Charts for Your Website," Chartjs.org, 2019, `https://www.chartjs.org/` (accessed Dec. 23, 2023).

5. Google, "Google Fonts," Google Fonts, 2019, `https://fonts.google.com/` (accessed Dec. 23, 2023).

6. Font Awesome 5, "Font Awesome 5," Fontawesome.com, 2017, `https://fontawesome.com/` (accessed Dec. 23, 2023).

7. M. Otto, "Bootstrap," Bootstrap, 2022, `https://getbootstrap.com/` (accessed Dec. 23, 2023).

8. "bootstrap," cdnjs, `https://cdnjs.com/libraries/bootstrap/5.0.0` (accessed Mar. 26, 2024).

9. "Chart.js," cdnjs, `https://cdnjs.com/libraries/Chart.js/3.8.0` (accessed Mar. 26, 2024).

10. Node.js, "HTTP | Node.js v18.8.0 Documentation," nodejs.org, `https://nodejs.org/api/http.html` (accessed Dec. 23, 2023).

11. Node.js, "File System | Node.js v18.1.0 Documentation," nodejs.org, `https://nodejs.org/api/fs.html` (accessed Dec. 23, 2023).

12. J. Richardson, "fs-extra," npm, Nov. 28, 2023, `https://www.npmjs.com/package/fs-extra` (accessed Feb. 11, 2024).

13. Axios, "axios," npm, Jan. 25, 2024, `https://www.npmjs.com/package//axios` (accessed Feb. 11, 2024).

14. Weka, "LibSVM," Mvn Repository, Nov. 20, 2016, `https://mvnrepository.com/artifact/nz.ac.waikato.cms.weka/LibSVM` (accessed Feb. 12, 2024).

15. S. Leary, "JSON In Java," Mvn Repository, May 22, 2020, `https://mvnrepository.com/artifact/org.json/json/20200518` (accessed Feb. 12, 2024).

CHAPTER 7

NTMA and VQA Integration

With the aim of bringing together network traffic monitoring and analysis (NTMA) and video quality assessment (VQA) in a seamless and real-time application that addresses both network and video quality monitoring requirements, this chapter provides the full system model that bridges the gap between network activity analysis and video quality evaluation. The JavaScript codes and changes made to the Node.js server to accommodate real-time communication between the client and itself are thus presented, together with a complete program structure and implementation method. This is concluded with an application testing section that sheds light on the effectiveness of the combined NTMA and VQA application.

7.1 System Model for Integrated NTMA and VQA Application

In this section, the network traffic monitoring and analysis (NTMA) and video quality assessment (VQA) applications are integrated into a single seamless application. The same pieces of code are modified to create a three-layer architecture consisting of a client, backend server on Node.js, and local Java servlet hosted using Apache Tomcat. The three frameworks are run on the same device, but can be adapted for operating on different devices. With NTMA and VQA at its core, this application provides the end user with the following capabilities after performing a comprehensive assessment of the overall user experience:

1. Monitor real-time network parameters directly in the web browser through a browser extension.

2. Obtain regression and classification analytics results based on historical data collated into real-time values.

© Tulsi Pawan Fowdur, Lavesh Babooram 2024
T. P. Fowdur, L. Babooram, *Machine Learning For Network Traffic and Video Quality Analysis*,
https://doi.org/10.1007/979-8-8688-0354-3_7

3. Obtain a Quality of Service (QoS) score of the network performance, referred to as the network score.

4. Monitor real-time video quality metrics directly in the web browser through a browser extension.

5. Obtain a live mean opinion score (MOS) of the video stream being watched.

6. Obtain a dominant quality rating summarizing the video quality performance over the frames analyzed.

The principles making up this application are inherited from the NTMA and VQA applications in Chapter 5 and Chapter 6, respectively. The user interface (UI) is thus built to display a blend of metrics from the two realms and real-time graphs that can be viewed dynamically via a toggle selection between live NTMA and live VQA, giving the end user complete control over viewing preferences. As explained in the previous chapters, the same four fundamental files are needed for the extension to function. This complete system model is given in Figure 7-1.

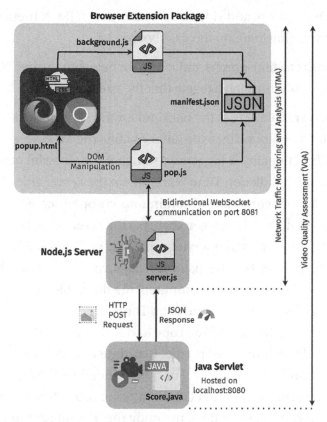

Figure 7-1. *Complete Integrated NTMA and VQA system model*

7.1.1 Components and Functionalities

The same four files listed next make up the client side of this application, i.e., the frontend:

- popup.html

- background.js

- pop.js

- manifest.json

They serve the same functionalities as described in Tables 5-1 and 6-1, as listed here:

- Take user inputs from the end user for the "Window Size" and "Prediction Time" parameters.

- Trigger the Node.js server by sending a message to it when the "Monitor & Analyze" button is clicked.

- Receive the network and video quality metrics in JSON messages before outputting them on the dashboard.

- Update the real-time graphs and alternate between the live NTMA and VQA graphs through a toggle that the user can click.

Likewise, the server.js file acts as the backend for the NTMA block and the intermediary for the VQA block, since the latter obtains the MOS score from the Java servlet. The NTMA functionality is triggered through the successful reception of the "Window Size" and "Prediction Time" variables at the server side. On the same wavelength, the VQA block consisting of continuous cropped screenshots undergoing image quality assessment (IQA) is triggered once a full screenshot is reconstructed and cropped at the server side. The same servlet explained in Chapter 6 is then hosted to calculate the MOS score based on the distortion aggravation method. Once both the network and video quality metrics are computed, the dashboard gives an indication of the experience being observed by the end user. Therefore, when a user visits a video streaming website among the likes of YouTube and Twitch, the integrated browser extension can be used to monitor and predict real-time network metrics, as well as to observe the evolution in video quality. For this application, the NTMA monitoring process is set to sixty seconds, and the VQA monitoring procedure to sixty frames. The latter involves more processing time, meaning that the full, complete test is only available after obtaining the VQA results.

7.1.2 Prediction and Classification of Network Traffic with Video Quality Metrics

The integrated NTMA/VQA application consists of the monitoring processes occurring seamlessly at the server side, while reporting real-time metrics and updates to the client side. Despite functioning independently, these two blocks can go hand-in-hand to improve granularity and insights. These metrics can serve as strong and reliable measures of the performance of a typical network undergoing different types of activities, and even experiencing spikes and dips. The more the machine learning (ML) algorithms are trained, the better the accuracy is, hence the anticipation and reactiveness of the algorithms to unexpected traffic behaviors. The key to the performance of this application lies in the dataset selection, depending on the networking environment. This framework provides the basis for building a reliable integrated NTMA and VQA application available through a simple click in a web browser.

7.1.3 Integrated NTMA/VQA Application Layout

In this application, the popup gives the user the ability to monitor real-time network and video metrics directly in the web browser. A pair of live graphs depict the evolution of the metrics in real time. A selection between the NTMA and VQA graphs is available through a toggle. Once opened, the client searches for the Node.js server and connects to it. The user clicks on the "Monitor & Analyze" button to trigger the backend processes. The same functionalities seen in Chapter 5 and Chapter 6 then occur, allowing the UI to be updated with real-time metrics, accommodated by the dashboard cards, which are categorized as "Network Metrics" or "Video Quality Metrics." The tests conclude with a blend of network and video metrics produced after the analysis of sixty instances for each block, thus making up a particular session. Figure 7-2 depicts the integrated NTMA/VQA browser extension, with both graphs displayed.

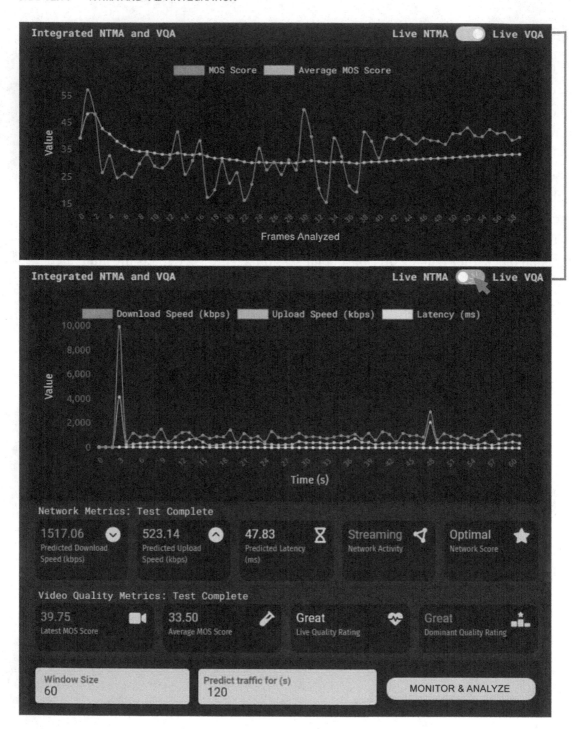

Figure 7-2. *Integrated NTMA and VQA application layout*

7.1.4 Client–Server–Servlet Interaction

Figure 7-3 shows the stepwise three-way communication architecture between the client, the Node.js server, and the hosted Apache Tomcat Java servlet.

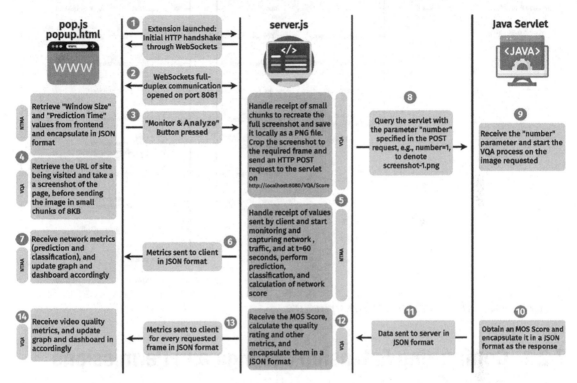

***Figure 7-3.** Client–Server–Servlet interaction for integrated NTMA and VQA*

7.2 Client Program Structure for Integrated NTMA/ VQA Application

This section contains the frontend development through the four basic files. An outline of the client-side program structure is shown in Figure 7-4.

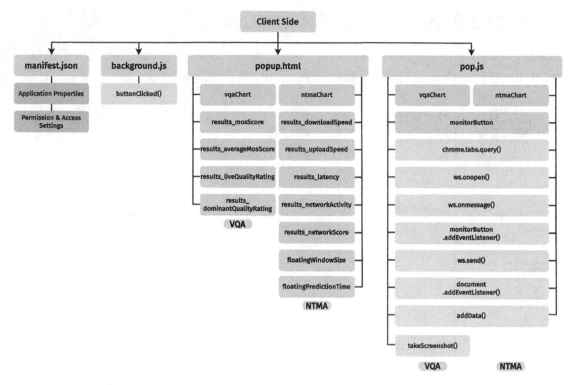

Figure 7-4. *Client-side program structure*

7.2.1 Configuring Extension Settings and Permissions

In this section, the manifest.json file containing metadata that describes the project and allows Google Chrome to add the application to the browser extension panel is given in Listing 7-1. The "icons" tag is either catered for by adding three icons of file type PNG to the directory folder of the project, or omitted in case the user wants the default icon to appear.

Listing 7-1. Configuring manifest.json with extension settings and permissions

```
{
  "manifest_version": 3,  // Manifest format version for Chrome extensions.

  "name": "Integrated NTMA and VQA",  // Name of the Chrome extension.
  "version": "1.0",  // Version number of the Chrome extension.

  "icons": {
    "16": "icon16.png",  // Icon for 16x16 pixel size.
```

```
    "48": "icon48.png",    // Icon for 48x48 pixel size.
    "128": "icon128.png"   // Icon for 128x128 pixel size.
  },

  "background": {
    "service_worker": "background.js"  // Background script, implemented as
    a service worker.
  },

  "permissions": [ // List of permissions required by the extension.
    "activeTab",   // Allows extension to interact with the currently
                      active tab.
    "scripting",   // Permits the injection of scripts into web pages.
    "webNavigation",  // Enables the extension to receive events in the web
                         navigation lifecycle.
    "webRequest",  // Grants the extension access to observe and analyze
                      network requests.
    "storage",   // Provides access to the browser's storage APIs for data
                    persistence.
    "declarativeNetRequest",  // Allows the extension to declare a set of
                                 rules for web requests.
    "declarativeNetRequestFeedback",  // Permits the extension to provide
                                         feedback on web requests.
    "tabs"  // Enables the extension to query and manipulate browser tabs.
  ],

  "action": {
    "default_popup": "popup.html",  // Default popup HTML file when the
                                       extension icon is clicked.
    "default_title": "Integrated NTMA and VQA"  // Default title for the
                                                    extension icon.
  }
}
```

7.2.2 Configuring the Background Script

Similar to the NTMA and VQA applications, the code for the background script, i.e., the background.js file, is given in Listing 7-2.

Listing 7-2. Configuring background.js with service worker logic

```
// Log a message indicating that the extension has been launched.
console.log("Extension launched...");

// Listen for the extension's action button click event.
chrome.action.onClicked.addListener(buttonClicked);

// Function to handle the button click event.
function buttonClicked(tab) {
  // Log a message indicating the background script is running when the
  button is clicked.
  console.log("Background script running...");
}

// Additional service worker logic can be added here
// This script operates separately from the extension and can perform
background tasks efficiently
// Developers can extend the functionality of the service worker to handle
various tasks, such as
// network requests, data management, and background processing.
```

7.2.3 Building the User Interface

The UI is built with HTML and CSS as the foundation to create a popup containing a toggle that switches between the NTMA and VQA real-time graphs when clicked. The dashboard also contains the network and video metrics rows, in which the dashboard cards are housed. This section thus provides the user with an informative dive into both network and video quality metrics.

File Functionality

The popup.html file is coded with the following functionalities:

1. Specify the document structure consisting of the following:

   i. Document type and language.

   ii. Charset and title.

   iii. Modify the content security policy (CSP) to allow certain script sources.

2. Add references to external resources:

   i. Import the required font stylesheets from Google Fonts [1].

   ii. Import icons for the dashboard from the Font Awesome icon library [2].

   iii. Link a custom Bootstrap stylesheet downloaded from Bootstrap's official website[3].

3. Add internal CSS for modifying the component appearance:

   i. Set styling for the body, background, and buttons.

   ii. Define styling for user interface components such as buttons, labels, and charts.

4. Display a title and a checkbox:

   i. Create a container for embedding the title and checkbox.

   ii. Give the popup a title.

   iii. Create a checkbox, which is formatted through CSS to appear like a toggle.

5. Display two chart areas:

   i. Create a container for embedding the NTMA chart area.

   ii. Create a container for embedding the VQA chart area.

   iii. Embed a chart canvas with Chart.js for each one.

6. Display the "Network Metrics" cards:

   i. Create a container for network metrics.

   ii. Display the predicted download speed, predicted upload speed, predicted latency, network activity, and network score individually along with their icons.

7. Display the "Video Quality Metrics" cards:

    i. Create a container for video quality metrics.

    ii. Display the latest MOS score, average MOS score, live quality rating, and dominant quality rating individually along with their icons.

8. Add fields for user inputs:

    i. Define input fields for window size and prediction time.

    ii. Define the "Monitor & Analyze" button.

9. Add downloaded and client scripts:

    i. Link the downloaded Chart.js library [4].

    ii. Link the custom pop.js script.

Libraries and Required Resources

A review of the listed libraries and downloadable scripts, along with their respective uses, is given in Table 7-1.

Table 7-1. *Description of Libraries and Resources for the User Interface*

Library/ Resource	Purpose	Repository	Download Link	Location
Google Fonts	Provides custom fonts for the web page	[1]		popup.html
Font Awesome	Provides scalable vector icons	[2]		popup.html
bootstrap.min. css	Customized Bootstrap stylesheet for organizing row and columns	[3]	[5]	"css" folder inside project's root folder
chart.js	Provides an interactive chart for the live graph.	[4]	[6]	Project's root folder
pop.js	Custom JS file for the logic	N/A		Project's root folder

Creating the Document Structure

The code given in Listing 7-3 is first used to create the page structure for popup.html. The placement of the remaining codes in this structure are also shown.

Listing 7-3. Creating the document structure

```
<!DOCTYPE html>
<html lang="en" dir="ltr">
   <head>
      <!-- Listing 7-4 is included here -->
      <style>
      <!-- Listing 7-6 is included here -->
      </style>
   </head>
   <body>
      <!-- Listing 7-5 is included here -->
   </body>
</html>
```

Adding the Document Details and References to External Resources

To add the fonts, icons, and styles, the code given in Listing 7-4 is added inside the head element.

Listing 7-4. Adding document details and references to external resources

```
<meta charset="utf-8">
<!-- Set Content Security Policy to allow scripts from the same origin
('self') and unsafe-eval for script execution -->
<meta http-equiv="Content-Security-Policy" content="script-src 'self'
'unsafe-eval'; object-src 'self'">
<!-- Title for the tab -->
<title>Integrated NTMA and VQA</title>
<!-- Font Stylesheet -->
<link rel="stylesheet" type="text/css" href="http://fonts.googleapis.com/cs
s?family=Roboto:regular,bold&subset=Latin">
<link rel="stylesheet" type="text/css" href="http://fonts.googleapis.com/
```

```
css?family=Fira Sans:regular,bold&subset=Latin">
<link rel="stylesheet" type="text/css" href="http://fonts.googleapis.com/
css?family=Roboto Mono:regular,bold&subset=Latin">
<!-- Icons Stylesheet -->
<link href="https://cdnjs.cloudflare.com/ajax/libs/font-awesome/6.4.2/css/
all.min.css" rel="stylesheet">
<!-- Customized Bootstrap Stylesheet -->
<!-- This file is downloaded from "https://getbootstrap.com/" online and
placed in the "css" folder. -->
<link href="css/bootstrap.min.css" rel="stylesheet">
```

Adding the Toggle, Graphs, and Dashboard Components

The graphs and dashboard components, together with the external scripts—namely
chart.js and pop.js—are added in the body tag. These scripts are placed at the end to
enable the document object model (DOM) to fully load before execution. This code is
given in Listing 7-5.

Listing 7-5. Adding the toggle, graphs, and dashboard components

```
<!-- Title and Toggle Container -->
<div class="title-toggle-container">
    <!-- Title to be displayed at the top of popup -->
    <p id="title" class="mb-0">Integrated NTMA and VQA</p>
    <!-- Toggle Container -->
    <div class="toggle-container">
        <!-- Toggle Captions -->
        <span class="toggle-caption">Live NTMA</span>
        <!-- Toggle Switch -->
        <label class="switch">
            <input type="checkbox">
            <span class="slider round"></span>
        </label>
        <!-- Toggle Captions -->
        <span class="toggle-caption">Live VQA</span>
    </div>
</div>
```

```html
<!-- Chart Area for NTMA -->
<div class="container-fluid pt-2" id="ntmaGraph">
    <div class="row g-2">
        <div class="col-sm-12 col-xl-3">
            <!-- Container for the chart -->
            <div class="bg-secondary rounded d-flex align-items-center
            justify-content-between p-3" style="margin-
            bottom: 2px;">
                <!-- Canvas for the chart -->
                <canvas id="ntmaChart" height="250" width="646"></canvas>
            </div>
        </div>
    </div>
</div>

<!-- Chart Area for VQA -->
<div class="container-fluid pt-2" id="vqaGraph">
    <div class="row g-2">
        <div class="col-sm-12 col-xl-3">
            <!-- Container for the chart -->
            <div class="bg-secondary rounded d-flex align-items-center
            justify-content-between p-3" style="margin-
            bottom: 2px;">
                <!-- Canvas for the chart -->
                <canvas id="vqaChart" height="250" width="646"></canvas>
            </div>
        </div>
    </div>
</div>

<!-- Dashboard Components - Network Metrics Cards -->
<div class="dashboard-container">
    <!-- Title for the Network Metrics -->
    <p class="dashboard-title" id="results_network">Network Metrics</p>
    <!-- Row for the Network Metrics Cards -->
```

```
<div class="dashboard-row">
    <!-- Card for Download Speed -->
    <div class="dashboard-card">
        <!-- Icon for Download Speed -->
        <i class="icon fas fa-circle-chevron-down"></i>
        <!-- Download Speed Metric to be displayed -->
        <span class="metric" id="results_downloadSpeed" style="color:
        #FF6384;">Hold on...</span>
        <!-- Description for Download Speed -->
        <p class="description">Predicted Download Speed (kbps)</p>
    </div>
    <!-- Card for Upload Speed -->
    <div class="dashboard-card">
        <!-- Icon for Upload Speed -->
        <i class="icon fas fa-circle-chevron-up"></i>
        <!-- Upload Speed Metric to be displayed -->
        <span class="metric" id="results_uploadSpeed" style="color:
        #2CD3E1;">Hold on...</span>
        <!-- Description for Upload Speed -->
        <p class="description">Predicted Upload Speed (kbps)</p>
    </div>
    <!-- Card for Latency -->
    <div class="dashboard-card">
        <!-- Icon for Latency -->
        <i class="icon fas fa-hourglass"></i>
        <!-- Latency Metric to be displayed -->
        <span class="metric" id="results_latency" style="color:
        #45FFCA;">Hold on...</span>
        <!-- Description for Latency -->
        <p class="description">Predicted Latency (ms)</p>
    </div>
    <!-- Card for Network Activity -->
    <div class="dashboard-card">
        <!-- Icon for Network Activity -->
        <i class="icon fas fa-circle-nodes"></i>
```

```
    <!-- Network Activity Metric to be displayed -->
    <span class="metric" id="results_networkActivity" style="color:
    #FF6384;">Hold on...</span>
    <!-- Description for Network Activity -->
    <p class="description">Network Activity</p>
</div>
<!-- Card for Network Score -->
<div class="dashboard-card">
    <!-- Icon for Network Score -->
    <i class="icon fas fa-star"></i>
    <!-- Network Score Metric to be displayed -->
    <span class="metric" id="results_networkScore" style="color:
    #2CD3E1;">Hold on...</span>
    <!-- Description for Network Score -->
    <p class="description">Network Score</p>
</div>
</div>
</div>

<!-- Dashboard Components - Video Quality Metrics Cards -->
<div class="dashboard-container" style="margin-top: 5px">
    <!-- Title for Video Quality Metrics -->
    <p class="dashboard-title" id="results_video">Video Quality Metrics</p>
    <!-- Row for the Video Quality Metrics Cards -->
    <div class="dashboard-row">
        <!-- Card for Latest MOS Score -->
        <div class="dashboard-card">
            <!-- Icon for Latest MOS Score -->
            <i class="icon fas fa-video"></i>
            <!-- Latest MOS Score Metric to be displayed -->
            <span class="metric" id="results_mosScore" style="color:
            #FF6384;">Hold on...</span>
            <!-- Description for Latest MOS Score -->
            <p class="description">Latest MOS Score</p>
        </div>
```

```
    <!-- Card for Average MOS Score -->
    <div class="dashboard-card">
        <!-- Icon for Average MOS Score -->
        <i class="icon fas fa-vial"></i>
        <!-- Average MOS Score Metric to be displayed -->
        <span class="metric" id="results_averageMosScore" style="color:
        #2CD3E1;">Hold on...</span>
        <!-- Description for Average MOS Score -->
        <p class="description">Average MOS Score</p>
    </div>
    <!-- Card for Live Video Quality Rating -->
    <div class="dashboard-card">
        <!-- Icon for Live Video Quality Rating -->
        <i class="icon fas fa-heart-pulse"></i>
        <!-- Live Video Quality Rating Metric to be displayed -->
        <span class="metric" id="results_liveQualityRating" style="color:
        #45FFCA;">Hold on...</span>
        <!-- Description for Live Video Quality Rating -->
        <p class="description">Live Quality Rating</p>
    </div>
    <!-- Card for Dominant Video Quality Rating -->
    <div class="dashboard-card">
        <!-- Icon for Dominant Video Quality Rating -->
        <i class="icon fas fa-ranking-star"></i>
        <!-- Dominant Video Quality Rating Metric to be displayed -->
        <span class="metric" id="results_dominantQualityRating"
        style="color: #FF4DC0;">Hold on...</span>
        <!-- Description for Dominant Video Quality Rating -->
        <p class="description">Dominant Quality Rating</p>
    </div>
    </div>
</div>

<!-- Fields for user inputs -->
<div class="container-fluid pt-1">
```

```
<!-- Row for user input fields -->
<div class="row g-2">
    <!-- Column for user input fields -->
    <div class="col-sm-12 col-xl-3">
        <!-- Background and styling for user input fields -->
        <div class="bg-secondary rounded d-flex align-items-center p-2"
        style="height: 75px">
            <!-- "Window Size" input field -->
            <div class="form-floating mb-1" style="padding-right: 5px;">
                <!-- Text input for "Window Size" -->
                <input type="text" class="form-control"
                id="floatingWindowSize" placeholder="" value="60">
                <!-- Label for "Window Size" -->
                <label for="floatingWindowSize" id="floatingLabels">Window
                Size</label>
            </div>
            <!-- "Prediction Time" input field -->
            <div class="form-floating mb-1" style="padding-left: 5px;">
                <!-- Text input for "Prediction Time" -->
                <input type="text" class="form-control"
                id="floatingPredictionTime" placeholder="" value="120"
                style="width: 240px;">
                <!-- Label for "Prediction Time" -->
                <label for="floatingPredictionTime"
                id="floatingLabels">Predict traffic for (s)</label>
            </div>
            <!-- Button for starting the monitoring and analysis processes,
            i.e., triggering the server -->
            <button id="monitorButton">Monitor & Analyze</button>
        </div>
    </div>
</div>
</div>
```

```
<!-- Include scripts for chart.js and pop.js at the end of the body to
ensure DOM is fully loaded before execution -->
<script src="chart.js"></script>
<script src="pop.js"></script>
```

Styling the Components through Internal CSS

To improve the user-friendliness and readability of the application, the code in
Listing 7-6 is added within the style tag.

Listing 7-6. Styling the components through internal CSS

```
/* Styles for the body */
body {
background-color: #131F33;
color: #92ABCF;
font-family: "Roboto";
font-size: 15px;
width: 710px;
}
/* Styles for the body's background */
.bg-secondary {
background-color: #0F1724 !important;
}
/* Styles for any primary text */
.text-primary {
color: #92ABCF !important;
}
/* Styles for the "Monitor & Analyze" Button */
#monitorButton {
display: flex;
width: 200px;
height: 30px;
align-content: right;
margin-top: 10px;
margin-bottom: 10px;
margin-left: 12px;
```

```css
text-transform: uppercase;
flex-direction: row;
-moz-box-align: center;
align-items: center;
gap: 0.375rem;
border: medium none;
box-shadow: rgba(0, 0, 0, 0.2) 0px 3px 1px -2px, rgba(0, 0, 0, 0.14) 0px
2px 2px 0px, rgba(0, 0, 0, 0.12) 0px 1px 5px 0px;
font-weight: bold;
font-family: Fira Sans, Arial, Helvetica, sans-serif;
cursor: pointer;
transition: background-color 250ms cubic-bezier(0.4, 0, 0.2, 1) 0ms,
box-shadow 250ms cubic-bezier(0.4, 0, 0.2, 1) 0ms, border 250ms cubic-
bezier(0.4, 0, 0.2, 1) 0ms;
-moz-box-pack: center;
justify-content: center;
border-radius: 12px;
padding: 0.625rem 0.75rem;
font-size: 0.820rem;
color: rgb(0, 0, 0);
background-color: rgb(17, 236, 229);
}
/* Styles for appearance of button upon hover */
#monitorButton:hover {
background-color: #bd1515;
color: white;
transition: .5s;
}
/* Styles for "Window Size" field */
#floatingWindowSize {
background-color: rgb(17, 236, 229);
color: rgb(0, 0, 0);
height: 50px;
}
```

```
/* Styles for appearance of the "Window Size" field when clicked */
#floatingWindowSize:focus {
border: 3px solid rgb(17, 236, 229) !important;
box-shadow: 0 0 3px rgb(17, 236, 229) !important;
-moz-box-shadow: 0 0 3px rgb(17, 236, 229) !important;
-webkit-box-shadow: 0 0 3px rgb(17, 236, 229) !important;
}
/* Styles for "Predict traffic for (s)" field */
#floatingPredictionTime {
background-color: rgb(17, 236, 229);
color: rgb(0, 0, 0);
height: 50px;
}
/* Styles for appearance of the "Predict traffic for (s)" field when
clicked */
#floatingPredictionTime:focus {
border: 3px solid rgb(17, 236, 229) !important;
box-shadow: 0 0 3px rgb(17, 236, 229) !important;
-moz-box-shadow: 0 0 3px rgb(17, 236, 229) !important;
-webkit-box-shadow: 0 0 3px rgb(17, 236, 229) !important;
}
/* Styles for "Window Size" and "Predict traffic for (s)" fields */
#floatingLabels {
color: rgb(0, 0, 0);
font-weight: bold;
}
/* Styles for any input fields */
input:focus, textarea:focus, select:focus {
outline-offset: 0px !important;
outline: none !important;
}
/* Styles for the title of the popup */
#title {
padding-top: 2px;
padding-left: 15px;
```

```
text-align: left;
font-family: Roboto Mono;
font-weight: bold;
font-size: 14px;
color: #DFCCFB;
}
/* Styles for the title of each dashboard row */
.dashboard-title {
color: #DFCCFB;
margin-left: 10px;
margin-bottom: -5px;
font-size: 13px;
font-family: 'Roboto Mono';
}
/* Styles for a dashboard row */
.dashboard-row {
display: flex;
justify-content: space-between;
}
/* Styles for a dashboard card */
.dashboard-card {
padding: 7px;
margin: 6px;
border-radius: 10px;
background-color: #101424;
position: relative;
width: 225px;
}
/* Styles for a dashboard container */
.dashboard-container {
background-color: #1d243f;
margin-top: 2px;
margin-left: 10px;
padding-left: 5px;
border-radius: 10px;
position: relative;
```

```css
width: 97%;
}
/* Styles for a dashboard icon */
.icon {
position: absolute;
top: 10px;
right: 10px;
font-size: 20px;
color: #DFCCFB;
}
/* Styles for a dashboard metric */
.metric {
font-size: 16px;
}
/* Styles for a dashboard description */
.description {
font-size: 10px;
font-family: Fira Sans;
margin-bottom: 12px;
margin-right: 16px;
}
/* Styles for the title and toggle container */
.title-toggle-container {
display: flex;
align-items: center;
justify-content: space-between; /* Align items to the far right */
font-family: Roboto Mono;
font-weight: bold;
font-size: 14px;
color: #DFCCFB;
}
/* Styles for the toggle container */
.toggle-container {
padding-top: 2px;
display: flex;
```

```css
align-items: center;
}
/* Styles for the toggle captions */
.toggle-caption {
font-size: 14px;
color: #DFCCFB;
margin-left: 10px;
margin-right: 10px;
}
/* Styles for the toggle switch */
.switch {
position: relative;
display: inline-block;
width: 40px;
height: 20px;
}
/* Styles for the toggle input */
.switch input {
opacity: 0;
width: 0;
height: 0;
}
/* Styles for the slider (the toggle button) */
.slider {
position: absolute;
cursor: pointer;
top: 0;
left: 0;
right: 0;
bottom: 0;
background-color: #5FBDFF;
-webkit-transition: .4s;
transition: .4s;
border-radius: 34px;
}
```

```css
/* Styles for the slider's thumb */
.slider:before {
position: absolute;
content: "";
height: 14px;
width: 14px;
left: 3px;
bottom: 3px;
background-color: white;
-webkit-transition: .4s;
transition: .4s;
border-radius: 50%;
}
/* Styles for the slider when input is checked */
input:checked + .slider {
background-color: #15F5BA; /* Changes background color when checked */
}
/* Styles for the slider when input is focused */
input:focus + .slider {
box-shadow: 0 0 1px #2196F3; /* Adds box shadow when focused */
}
/* Styles for the slider's thumb when input is checked */
input:checked + .slider:before {
-webkit-transform: translateX(20px);
-ms-transform: translateX(20px);
transform: translateX(20px);
}
```

Adding the User Interface to Google Chrome

To add the UI to Google Chrome, the same steps given in Sections 5.2.3.7 and 6.2.3.7 can be followed. Once added, the integrated NTMA and VQA application is listed in the Extensions menu, as shown in Figure 7-5. It can also be pinned on the top right of the web browser by following the same steps discussed earlier.

Figure 7-5. *Integrated NTMA and VQA application added to the Extensions menu*

Visualizing the User Interface

Figure 7-6 gives the completed UI once the application is clicked in the quick access bar. For this application, since the appearance of the two charts is handled with JavaScript, i.e., in the pop.js file, this UI is obtained after building the client script.

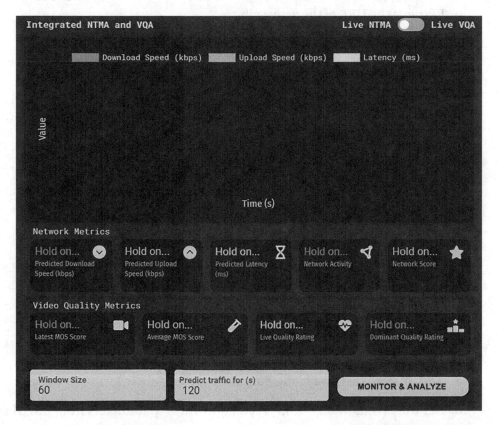

Figure 7-6. *Integrated NTMA and VQA application completed user interface*

7.2.4 Building the Client Script

The client's functionality is implemented in JavaScript and housed in a single file denoted as pop.js. As shown in the previous NTMA and VQA examples, this file manages communication to and from the Node.js server using WebSockets [7], refreshes the UI with the received metrics, and updates the two graphs with their respective metrics by keeping track of the index reached using Chart.js.

File Functionality

The pop.js file has the following functions:

1. Declare global variables for use throughout the script.

2. Query the Chrome tab using Chrome extension API:

   a. Retrieve information about the active and last focused tab.

   b. Retrieve the URL of the active tab.

3. Create a WebSocket connection to the server:

   a. Implement method for receiving messages from the server.

   b. Handle the connection using the open event.

   c. Parse and log messages.

4. Receive and handle messages from the server:

   a. Parse JSON messages containing real-time network metrics, video quality metrics, and graph updates.

   b. Update HTML elements with the network and video quality metrics.

   c. Update the Chart.js chart dynamically with the latest metrics.

   d. Generate an informative alert based on the specific error sent back by the server.

5. Create an event listener for the "Monitor" button:

   a. Add an event listener to the button.

   b. Get the "Window Size" and "Prediction Time" parameters from the UI and send them to the server in a JSON format.

c. Call a method that takes a screenshot of the current page view using the Chrome extension API.

d. Calculate the number of 8 KB chunks it would take to send the image to the server.

e. Send the number of chunks the server should expect as metadata.

f. Send each chunk to the server.

6. Create an event listener for the DOM:

a. Add an event listener to the DOM.

b. Initialize two Chart.js chart objects for NTMA and VQA.

c. Add an event listener to the toggle.

d. Hide the NTMA graph and unhide the VQA graph when the toggle is checked.

e. Unhide the NTMA graph and hide the VQA graph when the toggle is unchecked.

f. Create configurations for the five datasets.

7. Implement a method to update the charts every second based on a specified label.

8. Implement a method to take a screenshot of the current page upon request from the Node.js server.

Creating the Script Structure

The structure of the pop.js script is given in Listing 7-7, together with the placement of code for other functionalities.

Listing 7-7. Creating the script structure

```
// Declare the global objects which are called throughout the script.
var monitorButton; // Variable for the "Monitor & Analyze" object.
var ntmaChart; // Variable for the NTMA Chart.js object.
var vqaChart; // Variable for the VQA Chart.js object.
```

```
// Query the active and last focused tab to retrieve the current URL.
chrome.tabs.query({
    active: true,
    lastFocusedWindow: true,
},
    function(tabs) {
        <!-- Listing 7-9 is included here -->
    }
);

// Adding an event listener on the DOM to execute as soon as the
page loads.
document.addEventListener('DOMContentLoaded', function() {

        <!-- Listing 7-8 is included here -->

}, false);

// Method to update the chart dynamically.
function addData(chart, label, data, updateLabel) {

        <!-- Listing 7-10 is included here -->

}
```

Initializing and Configuring the Charts on Page Load

As soon as the HTML document's DOM is loaded, the DOMContentLoaded event is triggered to execute its contents. The two graphs, ntmaGraph and vqaGraph, are thus references into chart1 and chart2, respectively, before the following operations take place:

1. Get the references to the two chart canvases for NTMA and VQA.

2. Get the reference to the toggle switch input element.

3. Hide the VQA chart initially.

4. Create a function to toggle between the two charts.

5. Get the DOM element for the "Monitor" button and assign it to the global variable monitorButton.

6. Create dataset configurations:

    i. Create three separate datasets for "Download Speed (kbps)," "Upload Speed (kbps)," and "Latency (ms)." These are for the NTMA graph.

    ii. Create two separate datasets for "MOS Score" and "Average MOS Score." These are for the VQA graph.

    iii. For each dataset, specify the label, background color, border color, border width, and point radius.

7. Create the `dataNTMA` and `dataVQA` objects, which encapsulate all the dataset settings, and thereby are fed directly as a parameter to Chart.js.

8. Create the `optionsNTMA` and `optionsVQA` objects, which style the appearance of the Chart.js chart, with settings including responsiveness, aspect ratio, line tension, axis configurations, tick display settings, and legend styles.

9. Create the `configNTMA` and `configVQA` objects for the Chart.js chart where the type of the chart, the data to be included, and the specified options are all encapsulated.

10. Create new Chart.js instances using the `ntmaChart` and `vqaChart` identifiers, thus binding all the settings to the HTML elements. This indicates the object where the line chart is rendered.

The code for the described functions is given in Listing 7-8.

Listing 7-8. Initializing and configuring the charts on page load

```
// Get references to the two chart canvases for NTMA and VQA.
var chart1 = document.getElementById('ntmaGraph');
var chart2 = document.getElementById('vqaGraph');

// Get reference to the toggle switch input element.
var toggleSwitch = document.querySelector('.switch
input[type-"checkbox"]');

// Hide the VQA chart initially.
chart2.style.display = 'none';
```

```javascript
// Function to toggle between the two charts.
toggleSwitch.addEventListener('change', function () {
    if (toggleSwitch.checked) {
        // When the toggle is checked, the VQA graph appears.
        chart1.style.display = 'none'; // Hide the NTMA graph.
        chart2.style.display = 'block'; // Un-hide the VQA graph.
    } else {
        // When the toggle is unchecked, the VQA graph appears.
        chart1.style.display = 'block'; // Un-hide the NTMA graph.
        chart2.style.display = 'none'; // Hide the VQA graph.
    }
});

// Get the DOM element for the "Monitor & Analyze" button.
monitorButton = document.getElementById('monitorButton');

// Create the dataset configuration for the "Download Speed" line chart.
const dataset_DS = {
    label: 'Download Speed (kbps)',
    backgroundColor: 'rgb(255, 99, 132)',
    borderColor: 'rgb(255, 99, 132)',
    borderWidth: 1.2,
    pointRadius: 1.6
};

// Create the dataset configuration for the "Upload Speed" line chart.
const dataset_US = {
    label: 'Upload Speed (kbps)',
    backgroundColor: 'rgb(44, 211, 225)',
    borderColor: 'rgb(44, 211, 225)',
    borderWidth: 1.2,
    pointRadius: 1.6
};

// Create the dataset configuration for the "Latency" line chart.
const dataset_Lat = {
    label: 'Latency (ms)',
    backgroundColor: 'rgb(69, 255, 202)',
```

```
        borderColor: 'rgb(69, 255, 202)',
        borderWidth: 1.2,
        pointRadius: 1.6
};

// Create the dataset configuration for the "MOS Score" line chart.
const dataset_MOS = {
        label: 'MOS Score',
        backgroundColor: 'rgb(255, 99, 132)',
        borderColor: 'rgb(255, 99, 132)',
        borderWidth: 1.2,
        pointRadius: 1.6
};

// Create the dataset configuration for the "Average MOS Score" line chart.
const dataset_AvgMOS = {
        label: 'Average MOS Score',
        backgroundColor: 'rgb(44, 211, 225)',
        borderColor: 'rgb(44, 211, 225)',
        borderWidth: 1.2,
        pointRadius: 1.6
};

// Create the data object which encapsulates all datasets for NTMA
into one.
const dataNTMA = {
        datasets: [dataset_DS, dataset_US, dataset_Lat]
};

// Create the data object which encapsulates all datasets for VQA into one.
const dataVQA = {
        datasets: [dataset_MOS, dataset_AvgMOS]
};

// Create the options object which is responsible for styling the graph's
appearance for NTMA.
const optionsNTMA = {
```

```
// Set the width to be responsive to the area allocated.
responsive: false,
// Disallow the aspect ratio to change.
maintainAspectRatio: false,
elements: {
    line: {
        tension: 0.35, // Add a slight curvature to the line graphs.
    },
},
// Modify x-axis configurations.
scales: {
    x: {
        display: true, // Show x-axis.
        title: {
            display: true, // Show title of x-axis.
            text: 'Time (s)', // Change label of x-axis.
            color: '#DFCCFB', // Change color of x-axis label.
            font: {
                family: 'Fira Sans', // Change font of x-axis label.
                size: 14, // Change font size of x-axis label.
            },
        },
        ticks: {
            display: true, // Show ticks on x-axis.
            color: '#6b7e9b', // Change label color of x-axis tick.
            font: {
                family: 'Ubuntu', // Change font of x-axis tick.
                size: 10, // Change font size of x-axis tick.
            },
            stepSize: 10, // Change the step size for ticks on x-axis.
        },
    },
    y: {
        display: true, // Show y-axis.
        title: {
```

```
            display: true, // Show title of y-axis.
            text: 'Value', // Change label of y-axis.
            color: '#DFCCFB', // Change color of y-axis label.
            font: {
                family: 'Fira Sans', // Change font of y-axis label.
                size: 14, // Change font size of y-axis label.
            },
        },
        ticks: {
            display: false, // Disable ticks on y-axis.
            color: '#6b7e9b', // Change label color of y-axis tick.
            font: {
                family: 'Ubuntu', // Change font of y-axis tick.
                size: 13, // Change font size of y-axis tick.
            },
        },
    },
},
plugins: {
    legend: {
        labels: {
            color: '#DFCCFB', // Change legend label color.
            font: {
                family: 'Roboto Mono', // Change font of legend labels.
            }
        }
    }
}
};

// Create the options object, which is responsible for styling the graph's
appearance for VQA.
const optionsVQA = {
    // Set the width to be responsive to the area allocated.
    responsive: false,
```

```
// Disallow the aspect ratio to change.
maintainAspectRatio: false,
elements: {
    line: {
        tension: 0.35, // Add a slight curvature to the line graphs.
    },
},
// Modify x-axis configurations.
scales: {
    x: {
        display: true, // Show x-axis.
        title: {
            display: true, // Show title of x-axis.
            text: 'Frames Analysed', // Change label of x-axis.
            color: '#DFCCFB', // Change color of x-axis label.
            font: {
                family: 'Fira Sans', // Change font of x-axis label.
                size: 14, // Change font size of x-axis label.
            },
        },
        ticks: {
            display: true, // Show ticks on x-axis.
            color: '#6b7e9b', // Change label color of x-axis tick.
            font: {
                family: 'Ubuntu', // Change font of x-axis tick.
                size: 10, // Change font size of x-axis tick.
            },
            stepSize: 10, // Change the step size for ticks on x-axis.
        },
    },
    y: {
        display: true, // Show y-axis.
        title: {
            display: true, // Show title of y-axis.
            text: 'Value', // Change label of y-axis.
            color: '#DFCCFB', // Change color of y-axis label.
```

```
            font: {
                family: 'Fira Sans', // Change font of y-axis label.
                size: 14, // Change font size of y-axis label.
            },
        },
        ticks: {
            display: false, // Disable ticks on y-axis.
            color: '#6b7e9b', // Change label color of y-axis tick.
            font: {
                family: 'Ubuntu', // Change font of y-axis tick.
                size: 13, // Change font size of y-axis tick.
            },
        },
    },
    },
    plugins: {
        legend: {
            labels: {
                color: '#DFCCFB', // Change legend label color.
                font: {
                    family: 'Roboto Mono', // Change font of legend labels.
                }
            }
        }
    }
};

// Configuration object for the line chart for NTMA.
const configNTMA = {
    type: 'line', // Set the chart type.
    data: dataNTMA, // Set the data for the chart.
    options: optionsNTMA // Set the options for styling and customization.
};
```

```
// Configuration object for the line chart for VQA.
const configVQA = {
    type: 'line', // Set the chart type.
    data: dataVQA, // Set the data for the chart.
    options: optionsVQA // Set the options for styling and customization.
};

// Create a new Chart instance using the configuration and attach it to the
canvas element with the id "ntmaChart".
ntmaChart = new Chart(document.getElementById('ntmaChart'), configNTMA);
// Create a new Chart instance using the configuration and attach it to the
canvas element with the id "vqaChart".
vqaChart = new Chart(document.getElementById('vqaChart'), configVQA);
```

Real-Time WebSocket Communication and Dynamic Graph Updates

With the same principles as for the NTMA and VQA block, the client is mainly responsible for making a connection with the server on a predefined port, followed by the exchange of messages over the established connection. The following functions are involved:

1. Get the URL of the active tab and store it in a variable.

2. Initialize a time variable, which acts as a time counter for the graph.

3. Initialize a frame variable, which keeps track of the number of frames being analyzed.

4. Create constants to dynamically differentiate between the two graphs.

5. Establish a WebSocket connection on "ws://localhost:8081/", i.e., through port 8081, for real-time communication.

   i. To send a message, the ws.open event is used. In this case, the URL of the active tab stored earlier is sent to the server for it to know which site is being visited.

ii. To receive a message, the `ws.onmessage` event is used. This method is automatically triggered upon receipt of any message from the server. The latter is programmed to request for another screenshot once it receives the results from the servlet, which it relays to the client. Moreover, it is also triggered every second since the server is programmed to send real-time network metrics to the client every second. A series of `if` statements are then used to filter the messages based on the required updates. The HTML elements are updated accordingly.

6. Check the receipt of a message related to any error occurred at the server side and generate an alert if so.

7. Trigger real-time graph updates using the `addData()` method, which takes as input the chart object, the timestep, the real-time value, and a tag for dictating which dataset is being populated among the five.

8. Create an event listener for the "Monitor & Analyze" button:

i. Encapsulate "Window Size" and "Prediction Time" variables, which are fetched from their respective HTML elements, into a JSON format using the `stringify()` method.

ii. Call a method that takes a screenshot of the current page view using the Chrome extension API.

iii. Implement the `takeScreenshot()` method outside of the "Monitor & Analyze" button's event listener.

The codes for these processes are given in Listing 7-9.

Listing 7-9. Real-time WebSocket communication and dynamic graph updates

```
// Store the URL of the active tab in a variable.
var tempurl = tabs[0].url;

// Variable for iterating time on the graph (NTMA).
var time = 0;
// Variable to keep track of the frames being analyzed (VQA).
var frame = 0;
```

```javascript
// Constants for updating graph data.
const updateDownload = "downloadTraffic";
const updateUpload = "uploadTraffic";
const updateLatency = "latency";
const updateMOS = "MOS";
const updateAverageMOS = "AverageMOS";

// Create a WebSocket connection to the server.
const ws = new WebSocket('ws://localhost:8081/');

// Handle the WebSocket connection "open" event.
ws.onopen = function() {
    // Log a message for when the extension has successfully connected to
    // the server.
    console.log('WebSocket Client Connected');

    // Send the URL of the active tab to the server for it to know which
    // site is being visited.
    // Specify the type as "sitedata" and send the message in a
    // JSON format.
    ws.send(JSON.stringify({type:'sitedata', value:tempurl}));
};

// Handle received messages from the server.
ws.onmessage = function(e) {
    // Log any received message.
    console.log("Received from server: " + e.data);

    // Handle messages related to Network Metrics.
    // The following statement checks whether the message received contains
    // the real-time or predicted values.
    // These object names are set at the server-side.
    if (e.data.includes("downloadSpeed") || e.data.includes
    ("uploadSpeed") || e.data.includes("latency") || e.data.
    includes("predictedDownloadSpeed") || e.data.includes("predicted
    UploadSpeed") || e.data.includes("predictedLatency")) {
        // Parse the JSON message for real-time metrics and graph updates.
        const msg = JSON.parse(e.data);
```

412

```
// Extract the predicted values from the message using the same
notation set at the server-side.
// Store these values in their corresponding variables.
var predRX = msg.predictedDownloadSpeed; // Variable for the
predicted Download Speed metric.
var predTX = msg.predictedUploadSpeed; // Variable for the
predicted Upload Speed metric.
var predLat = msg.predictedLatency; // Variable for the predicted
Latency metric.
var scoreQoS = msg.predictedQoS; // Variable for the QoS Score.
var deviceActivity = msg.deviceActivity; // Variable for the
classification results, i.e., the device activity.
var averageLatency = msg.averageLatency; // Variable for the
average latency metric (not displayed on the UI, but can be added
as a dashboard card).

// Check if the variables are not "undefined" since the values on
the UI are only updated after the results are obtained.
if(typeof predRX !== "undefined" || typeof predTX !== "undefined"
|| typeof predLat !== "undefined" || typeof scoreQoS !==
"undefined" || typeof deviceActivity !== "undefined") {
    // Update the HTML elements with the received metrics from
    the server.
    document.getElementById("results_downloadSpeed").innerHTML
    = predRX;
    document.getElementById("results_uploadSpeed").innerHTML
    = predTX;
    document.getElementById("results_latency").
    innerHTML = predLat;
    document.getElementById("results_networkScore").innerHTML =
    scoreQoS;
    document.getElementById("results_networkActivity").innerHTML =
    deviceActivity;
    // Update the UI to let the user know that the test is
    complete.
```

```
        document.getElementById("results_network").
        innerHTML = "Network Metrics: Test Complete";
    }

    // Extract thje real-time network metrics and graph updates.
    var RX = msg.downloadSpeed; // Store the real-time Download Speed
    metric in a variable.
    var TX = msg.uploadSpeed; // Store the real-time Upload Speed
    metric in a variable.
    var latency = msg.latency; // Store the real-time Latency metric in
    a variable.

    // Add data to the NTMA chart using the addData() method by parsing
    the following parameters:
    // 1. Chart object.
    // 2. Timestep (relative to when the user clicks on the button -
    this denotes t = 0).
    // 3. Real-time value.
    // 4. Tag for guiding the variable to the right set of values on
    the chart.
    addData(ntmaChart, time, RX, updateDownload); // Update the graph
    with the latest Download Speed variable.
    addData(ntmaChart, time, TX, updateUpload); // Update the graph
    with the latest Upload Speed variable.
    addData(ntmaChart, time, latency, updateLatency); // Update the
    graph with the latest Latency variable.

    // Increment time for the next data point.
    time = time + 1;
}

// Handle messages related to Video Quality metrics.
// The following statement checks whether the message received contains
the real-time or assessed video quality metrics.
// These object names are set at the server-side.
if (e.data.includes("videoScore") || e.data.includes("average
VideoScore") || e.data.includes("videoRating") || e.data.
includes("dominantRating")) {
```

```
// Parse the JSON message for real-time metrics and graph updates.
const msg = JSON.parse(e.data);

// Extract the predicted values from the message using the same
notation set at the server-side.
// Store these values in their corresponding variables.
var mosScore = parseFloat(msg.videoScore); // Variable for the
MOS score.
var averageMosScore = parseFloat(msg.averageVideoScore);
// Variable for the Average MOS score.
var liveQualityRating = msg.videoRating; // Variable for the Video
Quality Rating.
var dominantQualityRating = msg.dominantRating; // Variable for the
Dominant Video Quality Rating.

// Since this rating is live, update the HTML elements with the
received metrics from the server.
document.getElementById("results_liveQualityRating").innerHTML =
liveQualityRating;
document.getElementById("results_dominantQualityRating").innerHTML
= dominantQualityRating;

// Add data to the VQA chart using the addData() method by parsing
the following parameters:
// 1. Chart object.
// 2. Timestep (relative to when the result for a frame is being
received from the Java Servlet).
// 3. Real-time value.
// 4. Tag for guiding the variable to the right set of values on
the chart.
addData(vqaChart, frame, parseFloat(mosScore), updateMOS); //
Update the MOS graph with the information from the server.
addData(vqaChart, frame, parseFloat(averageMosScore),
updateAverageMOS); // Update the Average MOS graph with the
information from the server.
```

415

```
    // Increment the frame for the next frame to be analyzed.
    frame = frame + 1;

    // If the number of frames analyzed is under 60 (index starts at 0,
    so 0-59).
    if (e.data.includes("screenshot") && frame < 60) {
        // Call the method to take a screenshot.
        takeScreenshot();
    }

    // If the number of frames analyzed reaches 60.
    if (frame == 60) {
        // Update the HTML elements with the received metrics from
        the server.
        document.getElementById("results_mosScore").
        innerHTML = mosScore;
        // Round the average latency to two decimal places using the
        toFixed() method.
        document.getElementById("results_averageMosScore").innerHTML =
        averageMosScore.toFixed(2);
        // Update the UI to let the user know that the test is
        complete.
        document.getElementById("results_video").
        innerHTML = "Video Quality Metrics: Test Complete";
    }
}

// Handle messages related to any error sent by the server.
// The following statement checks whether the message received contains
the "error" keyword.
// This keyword is set at the server-side.
if (e.data.includes("No Connection")) {
    // Generate an alert to let the user know that latency values
    cannot be measured since there is no connection.
    alert("Unable to measure latency due to no connection.");
} else if (e.data.includes("Error cropping image.")) {
```

416

```
        // Generate an alert to let the user know that there was a problem
        with cropping the image.
        alert("Error cropping image.");
    } else if (e.data.includes("Error sending screenshot to servlet.")) {
        // Generate an alert to let the user know that there was a problem
        with sending the screenshot number to the servlet.
        alert("Error sending screenshot to servlet.");
    }
};

// Add an event listener to the "Monitor & Analyze" button.
monitorButton.addEventListener('click', function() {
    // Fetch the "window size" and "prediction time" variables from the UI.
    // These are sent to the server since they are needed for analytics.
    var wsize = document.getElementById('floatingWindowSize')
    .value;
    var ptime = document.getElementById('floatingPredictionTime')
    .value;

    // Send the variables for "window size" and "prediction time" to the
    server using the stringify() method.
    ws.send(JSON.stringify({type:'windowSize', value:wsize}));
    ws.send(JSON.stringify({type:'predictionTime', value:ptime}));

    // Call the method to take a screenshot.
    takeScreenshot();
}, false);

 // Method to capture a screenshot of the current page.
function takeScreenshot() {
    // Capture a screenshot.
    chrome.tabs.captureVisibleTab({ format: "png" }, function
    (screenshotUrl) {
        // Create a Blob from the data URL.
        fetch(screenshotUrl)
        .then((res) => res.blob())
        .then((blob) => {
```

```
        // Create a Promise to handle the conversion of Blob to
        ArrayBuffer.
        return new Promise((resolve) => {
            // Create a FileReader instance.
            const reader = new FileReader();
            // Resolve with the result when the reader finishes
            loading.
            reader.onloadend = () => resolve(reader.result);
            // Read the Blob as ArrayBuffer.
            reader.readAsArrayBuffer(blob);
        });
    })
    .then((arrayBuffer) => {
        // Specify a chunk size in which the image is broken down to be
        sent to the server.
        const chunkSize = 8192; // e.g., 8 KB chunks

        // Calculate the number of chunks needed.
        const totalChunks = Math.ceil(arrayBuffer.byteLength /
        chunkSize);

        // Send metadata first (total number of chunks).
        ws.send(JSON.stringify({type: "metadata", totalChunks}));

        // Chunk and send the data into as many chunks required.
        // Iterate through the chunks and send them over the WebSocket
        connection.
        for (let i = 0; i < totalChunks; i++) {
            // Calculate the start and end positions for the
            current chunk.
            const start = i * chunkSize;
            const end = Math.min((i + 1) * chunkSize, arrayBuffer.
            byteLength);
            // Slice the arrayBuffer to obtain the current chunk.
            const chunk = arrayBuffer.slice(start, end);
            // Send the chunk over the WebSocket connection.
            ws.send(chunk);
```

```
        }
    })
    .catch((error) => {
        // Log any errors in capturing the screenshot.
        console.error("Error capturing screenshot:", error);
    });
  });
}
```

Real-Time Chart Update Mechanism

This section relates to the addData() method, which is slightly modified using a combination of the contents from the NTMA and VQA blocks, where both graphs can be populated in real time. Its codes are given in Listing 7-10.

Listing 7-10. Real-time chart update mechanism

```
// Determine for which dataset the method has been called through the tag
previously made.
// If the tag represents Download Speed.
if (updateLabel === "downloadTraffic") {
    // Push the label, i.e., the time instant.
    chart.data.labels.push(label);
    // Push the data, i.e., the real-time metric.
    chart.data.datasets[0].data.push(data);
}
// If the tag represents Upload Speed.
else if (updateLabel === "uploadTraffic") {
    // Push the data, i.e., the real-time metric.
    chart.data.datasets[1].data.push(data);
}
// If the tag represents Latency.
else if (updateLabel === "latency") {
    // Push the data, i.e., the real-time metric.
    chart.data.datasets[2].data.push(data);
}
```

```
// If the tag represents the MOS.
if (updateLabel === "MOS") {
    // Push the label, i.e., the frame instant.
    chart.data.labels.push(label);
    // Push the data, i.e., the real-time metric.
    chart.data.datasets[0].data.push(data);
}
// If the tag represents the Average MOS.
else if (updateLabel === "AverageMOS") {
    // Push the data, i.e., the real-time metric.
    chart.data.datasets[1].data.push(data);
}

// Update the chart options to display the y-axis ticks when this method is
first called.
// This is done since the y-axis ticks are disabled at first.
ntmaChart.options.scales.y.ticks.display = true;
vqaChart.options.scales.y.ticks.display = true;

// Update the chart.
chart.update();
```

7.3 Server Program Structure for Integrated NTMA/VQA Application

The responsibilities of the `server.js` file include a combination of the processes detailed in Sections 5.3 and 6.3 where all modules are first imported, before initializing the global variables and creating an HTTP server. Then, the same mechanisms are used to handle WebSocket connections and accept the "Window Size" and "Prediction Time" parameters from the client, together with the reconstruction of the PNG image. This triggers the NTMA block, whose codes are the same, along with simultaneous communication with the Java servlet. The results for NTMA are obtained at the Node.js server end and are encapsulated in JSON format and sent to the client, while for VQA the Node.js server awaits the results from the servlet before doing the same. Logs are present at every checkpoint to verify the status of the server and provide insights into the flow of the procedures. Figure 7-7 outlines the components involved at the server side.

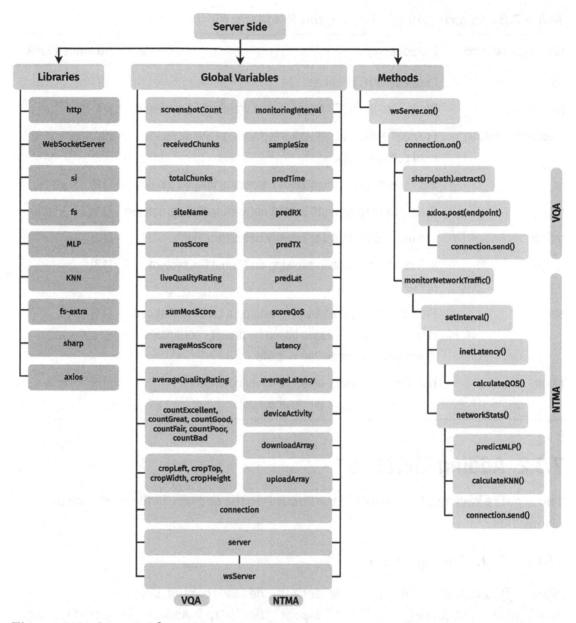

Figure 7-7. *Server-side program structure*

7.3.1 Libraries and Required Resources

Using the npm install keyword, all the libraries used for the NTMA and VQA applications are installed. Table 7-2 summarizes the libraries and resources required.

Table 7-2. *Description of Libraries and Resources for the Server*

Library/Resource	Purpose	Download Link
http	Creates the HTTP server	[8]
websocket	Binds onto the HTTP server for the WebSocket connection	[7]
systeminformation	Provides access to network interfaces on the device to read network traffic	[9]
fs	Works as a file system allowing access to files on the device	[10]
mlp	Creates a perceptron to predict network traffic parameters	[11]
ml-knn	Creates a KNN model to classify device activity	[12]
fs-extra	Works as a file system allowing additional file system operations	[13]
sharp	Processes images easily by providing a method that can crop the received images based on preset coordinates determined by the visited website	[14]
axios	Handles the request-response architecture for communicating with the Java servlet	[15]

7.3.2 Adding Libraries

The require keyword is used to add the libraries to the server.js file, as shown in Listing 7-11.

Listing 7-11. Adding libraries

```
//Specify imports related to the server and monitoring process.
const http = require('http'); // Import the 'http' module for creating an
                              HTTP server.
const WebSocketServer = require('websocket').server; // Import the
                                                       WebSocket
                                                       server module.
const si = require('systeminformation'); // Import the 'systeminformation'
                                          library for obtaining system
                                          information about the device.
```

```
const fs = require('fs'); // Import the 'fs' module for working with the
                              file system.
const fsExtra = require('fs-extra'); // Import the 'fs-extra' module for
                                        additional file system operations.
const sharp = require('sharp'); // Import the 'sharp' module for image
                                   processing.
const axios = require('axios'); // Import the 'axios' module for making
                                   HTTP requests.

// Specify imports related to Machine Learning.
var MLP = require('mlp'); // Import the 'mlp' library, which is used for
                             Multilayer Perceptron prediction.
var KNN = require('ml-knn'); // Import the 'ml-knn' library, which is used
                                for K-Nearest Neighbors classification.
```

7.3.3 Declaring Global Variables

A mixture of the global variables from the NTMA and VQA applications are added as shown in Listing 7-12.

Listing 7-12. Declaring global variables

```
// Specify imports related to Machine Learning.
var MLP = require('mlp'); // Import the 'mlp' library, which is used for
                             Multilayer Perceptron prediction.
var KNN = require('ml-knn'); // Import the 'ml-knn' library, which is used
                                for K-Nearest Neighbors classification.

// Declare and initialize the global variables to be used across the
server's script for NTMA.
var monitoringInterval; // Declare a variable for monitoring interval.
var sampleSize = 0; // Initialize a variable for sample size, use in the
                       data collection phase.
var predTime = 0; // Initialize a variable for prediction time at which the
                     user wants the prediction to be.
var predRX = 0; // Initialize a variable for prediction of download speed
                   in Kb/s.
```

423

```
var predTX = 0; // Initialize a variable for prediction of upload speed
                    in Kb/s.
var predLat = 0; // Initialize a variable for prediction of latency in ms.
var scoreQoS = ""; // Initialize a variable for the QoS score (Quality of
                    Service).
var latency = 0; // Initialize a variable for latency measurement in ms.
var averageLatency = 0; // Initialize a variable for calculating the
                    average latency in ms.
var deviceActivity = ""; // Initialize a variable for determining device
                    activity from "idle," "browsing," or "streaming."

// Declare and initialize the global variables to be used across the
server's script for VQA.
var screenshotCount = 1; // Counter for saving screenshots locally in
                    sequential order.
var receivedChunks = 0; // Counter for tracking received chunks of data.
var totalChunks; // Variable to store the total number of expected chunks.
var siteName = ""; // Variable to store the name of the website.
var mosScore = 0; // Variable to store the Mean Opinion Score (MOS).
var liveQualityRating = ""; // Variable to store the live quality rating.
var sumMosScore = 0; // Variable to accumulate the sum of MOS scores.
var averageMosScore = 0; // Variable to store the average MOS score.
var averageQualityRating = ""; // Variable to store the average
                    quality rating.
var countExcellent = 0; // Counter for the number of occurrences of
                    "Excellent" rating.
var countGreat = 0; // Counter for the number of occurrences of
                    "Great" rating.
var countGood = 0; // Counter for the number of occurrences of
                    "Good" rating.
var countFair = 0; // Counter for the number of occurrences of
                    "Fair" rating.
var countPoor = 0; // Counter for the number of occurrences of
                    "Poor" rating.
var countBad = 0; // Counter for the number of occurrences of "Bad" rating.
```

```
var cropLeft = 0; // Variable to store the left coordinate for
                     cropping images.
var cropTop = 0; // Variable to store the top coordinate for
                     cropping images.
var cropWidth = 0; // Variable to store the width for cropping images.
var cropHeight = 0; // Variable to store the height for cropping images.

// Define a variable to accept a connection from a client.
var connection;
```

7.3.4 Emptying the Screenshot Folders

The same codes from Section 6.3.4 are used to empty the respective screenshot folders, as given in Listing 7-13.

Listing 7-13. Emptying the screenshot folders

```
// Empty the "Screenshots" and "Screenshots-Cropped" folders.
// Please be careful which directory is being inserted here!
fsExtra.emptyDirSync("C://Users//Lavesh//Integrated//Screenshots");
fsExtra.emptyDirSync("C://Users//Lavesh//Integrated//ScreenshotsCropped");
```

7.3.5 Fetching the Local Databases

Likewise, the databases are fetched locally, through the codes in Listing 7-14, as explained in Section 5.3.4.

Listing 7-14. Fetching local databases

```
// Fetching some previously stored values from a locally stored database
for download speed.
const contents1 = fs.readFileSync("./Databases/downloadSpeed.txt",
'utf-8'); // Read the contents of the 'downloadSpeed.txt' file.
var downloadArray = contents1.split(/\r?\n/); // Split the contents into
                                                 an array, assuming each line
                                                 represents a value.
```

```
// Fetching some previously stored values from a locally stored database
for upload speed.
const contents2 = fs.readFileSync("./Databases/uploadSpeed.txt", 'utf-8');
// Read the contents of the 'uploadSpeed.txt' file.
var uploadArray = contents2.split(/\r?\n/); // Split the contents into
                                               an array, assuming each line
                                               represents a value.
```

7.3.6 Creating a WebSocket Server

The WebSocket server is hosted on port 8081 using the same aforementioned principles. This is shown in Listing 7-15.

Listing 7-15. Creating a WebSocket server

```
// Host a server connection on port 8081.
const server = http.createServer(); // Create an HTTP server instance.
server.listen(8081); // Listen on port 8081 for incoming HTTP requests.

console.log("Server started on port 8081..."); // Log a message indicating
that the server has started.
console.log("Waiting for connection..."); // Log a message indicating that
the server is waiting for connections.

// Create a WebSocket server instance and attach it to the HTTP server
previously created.
const wsServer = new WebSocketServer({
    httpServer: server,
});
```

7.3.7 Listening for a Client Connection

In this section, a connection from a requesting client is accepted, before waiting for a message to be received from it. At the client side, when the "Monitor & Analyze" button is clicked, the "Window Size" and "Prediction Time" variables are sent to the server. This occurs simultaneously with the screenshot-taking process. Now, these messages may arrive at the server side in a different order. For it to be able to differentiate between them, a "type" is assigned to each message at the client side. The five types of messages declared at the client side include the following:

- metadata: Denotes the reception of the number of chunks expected.

- sitedata: Denotes the URL of the site being visited.

- windowSize: Denotes the "Window Size" variable.

- predictionTime: Denotes the "Prediction Time" variable.

- binary: Denotes a chunk of the image screenshotted and sent by the client.

This section thus elaborates on the codes used to perform both NTMA and VQA at the server side.

Adding the VQA Block

The block of code that triggers the servlet upon the receipt and reconstruction of the full PNG image is given in Listing 7-16. It also invokes the monitorNetworkTraffic() method, which is explained in the next section.

Listing 7-16. Adding the VQA block

```
// Listen for a request for connection by a client through the
WebSocket server.
wsServer.on("request", function (request) {
        // Accept a connection from a client.
    connection = request.accept(null, request.origin);

    // Use the "message" keyword to accept incoming messages from
    the client.
    connection.on("message", function (message) {
        // Check if the received message type is in UTF-8 format.
        if (message.type === "utf8") {
            // Parse the UTF-8 data from the received message.
            const messageReceived = JSON.parse(message.utf8Data);
            // Check the type of the received message (metadata or
            sitedata).
```

```
        if (messageReceived.type === 'metadata') {
            // Update the totalChunks variable with the received
            metadata.
            totalChunks = messageReceived.totalChunks;
            // Log the received metadata information.
            console.log('Received metadata:', totalChunks, 'chunks
            expected.');
        } else if (messageReceived.type === 'sitedata') {
            // Update the siteName variable with the received value.
            siteName = messageReceived.value;
            // Log the received sitedata information.
            console.log('Received sitedata:', siteName, 'is the
            link.');
            // Check the website link to set cropping coordinates
            accordingly.
            // For YouTube and Twitch, the coordinates have been set.
            // For other websites, the whole page is taken as the
            screenshot for the VQA block.
            if (siteName.includes("youtube")){
                // For videos on https://www.youtube.com/
                cropLeft = 98;
                cropTop = 80;
                cropWidth = 1280;
                cropHeight = 720;
            } else if (siteName.includes("twitch")) {
                // For videos on https://www.twitch.tv/
                cropLeft = 239;
                cropTop = 52;
                cropWidth = 1340;
                cropHeight = 752;
            } else {
                // For videos on any other websites (no cropping).
                cropLeft = 0;
                cropTop = 0;
                cropWidth = 1920;
                cropHeight = 953;
```

```
        }
      } else if (messageReceived.type === 'windowSize') {
        // Check the type of the received message and if it
        corresponds to "windowSize".
        sampleSize = messageReceived.value; // Get the Window Size
        parameter.
      } else if (messageReceived.type === 'predictionTime') {
        // Check the type of the received message and if it
        corresponds to "predictionTime".
        predTime = messageReceived.value; // Get the Prediction
        Time parameter.
        // Both window size and prediction time values are
        obtained.
        // Call the method to trigger the NTMA process.
        monitorNetworkTraffic();
      }
    } else if (message.type === "binary") {
      // If the message received is of type binary.

      // Increment the counter for the number of chunks received.
      receivedChunks++;
      // Handle the binary data received.
      const arrayBuffer = message.binaryData;
      // Convert arrayBuffer to Buffer.
      const buffer = Buffer.from(arrayBuffer);

      // Append the chunk to a file (e.g., screenshot-1.png).
      // For one screenshot, each of its chunks is appended to the
      same file to create the final screenshot.
      fs.appendFileSync("C://Users//Lavesh//Integrated//Screenshots//
      screenshot-" + screenshotCount + ".png", buffer);

      // Log a message to indicate the index of chunks received.
      console.log('Received chunk:', receivedChunks, 'of',
      totalChunks);

      // If all chunks received, finalize and save the image file.
      if (receivedChunks === totalChunks) {
```

```
// This denotes that a full screenshot has been received
from the client side.
console.log("All chunks received. Screenshot saved.");

// Specify the paths of the file before and after cropping.
var pathBeforeCropping = "C://Users//Lavesh//Integrated//
Screenshots//screenshot-" + screenshotCount + ".png";
var pathAfterCropping = "C://Users//Lavesh//Integrated
//ScreenshotsCropped//screenshot-" + screenshotCount +
".png";

// Use the "Sharp" library to crop the image according to
the preset coordinates.
sharp(pathBeforeCropping)
  .extract({ left: cropLeft, top: cropTop, width:
  cropWidth, height: cropHeight })
  .toFile(pathAfterCropping, (err, info) => {
    if (err) {
        // Log an error message to indicate that cropping
        cannot be performed.
        console.error('Error cropping image:', err);
        // Create an error message to be sent over to
        the client.
        const errorMessage = {
            error: "Error cropping image."
        };
        // Encapsulate the error message in a JSON-
        formatted String and send it over the WebSocket
        connection.
        connection.send(JSON.stringify(errorMessage));
    } else {
        // Log a message to indicate that the image has
        been cropped successfully.
        console.log('Image cropped successfully:', info);

        // At this point, the cropped image containing the
        feed that the VQA block needs is available.
```

```
// Log a message indicating that a request is being
made to the Java HTTP Apache Server (servlet) to
return the MOS score.
console.log("Making request to servlet.")
// Use the "Axios" library to make an HTTP POST
request to the servlet.
axios.post("http://localhost:8080/VQA/
Score", null, {
    // Specify the parameter required by the
    servlet, in this case the screenshot count.
    params: {
        number: screenshotCount,
    },
})
.then((response) => {
    // Handle the response from the servlet.
    // Retrieve the MOS score from the JSON
    response.
    mosScore = response.data.score;
    mosScore = parseFloat(mosScore);
    // Log a message to indicate the MOS score
    obtained.
    console.log("Score obtained from servlet: " +
    mosScore);

    // Account for any abnormalities exceeding the
    ranges of excellent/bad quality.
    if (mosScore > 100) {
        mosScore = 100;
    } else if (mosScore < 0) {
        mosScore = 0;
    }
    // Map the MOS Score obtained to its quality
    rating for ease of understanding.
    // Increment the respective counter based on
    the quality assigned.
```

431

```
// This is to identify the most occuring video
quality, i.e., the "Dominant Quality Rating."
if (mosScore >= 0 && mosScore < 20) {
    liveQualityRating = "Excellent";
    countExcellent = countExcellent + 1;
}
else if (mosScore >= 20 && mosScore < 40) {
    liveQualityRating = "Great";
    countGreat = countGreat + 1;
}
else if (mosScore >= 40 && mosScore < 65) {
    liveQualityRating = "Good";
    countGood = countGood + 1;
}
else if (mosScore >= 65 && mosScore < 80) {
    liveQualityRating = "Fair";
    countFair = countFair + 1;
}
else if (mosScore >= 80 && mosScore < 90) {
    liveQualityRating = "Poor";
    countPoor = countPoor + 1;
}
else if (mosScore >= 90) {
    liveQualityRating = "Bad";
    countBad = countBad + 1;
}
// Log a message to indicate the live
quality rating.
console.log("Video Quality Rating: " +
liveQualityRating);

// Create an array to store the counts for each
quality rating.
var countQuality = [countExcellent, countGreat,
countGood, countFair, countPoor, countBad];
```

```
// Initialize variables to find the maximum
count and its index.
var maxCount = countQuality[0]; // Assume the
first one is the maximum.
var maxIndex = 0;
// Iterate through the countQuality array to
find the maximum count and its index.
for (var i = 1; i < countQuality.length; i++) {
    // Check if the current count is greater
    than the maximum count.
    if (countQuality[i] > maxCount) {
        // Update the maximum count and
        its index.
        maxCount = countQuality[i];
        maxIndex = i;
    }
}
// Create an array of labels representing the
quality ratings.
var labels = ['Excellent', 'Great', 'Good',
'Fair', 'Poor', 'Bad'];
// Determine the dominant quality rating based
on the maximum index.
var dominantQualityRating = labels[maxIndex];

// Log the dominant quality rating and its
count to the console.
console.log('The highest count label is:',
dominantQualityRating);
console.log('The highest count value is:',
maxCount);

// Calculate and log the Average MOS Score.
sumMosScore = sumMosScore + mosScore;
averageMosScore = sumMosScore /
screenshotCount;
```

```
            console.log("Average MOS Score: "
            +averageMosScore);

            // Send a message to the client to ask for
            another screenshot.
            // Create a constant variable holding the
            scores and ratings measured/calculated.
            const screenshotUpdate = {
                screenshot: "sendScreenshot",
                videoScore: mosScore,
                averageVideoScore: averageMosScore,
                videoRating: liveQualityRating,
                dominantRating: dominantQualityRating
            };

            // Encapsulate the metrics message in a JSON-
            formatted String and send it over the WebSocket
            connection.
            connection.send(JSON.stringify(screenshot
            Update));

            // Increment the screenshot count to move to
            the next one.
            screenshotCount = screenshotCount + 1;
        })
        .catch((error) => {
            // Log an error message to indicate that the
            screenshot cannot be sent to the servlet.
            console.error('Error sending screenshot to
            servlet:', error);
            // Create an error message to be sent over to
            the client.
            const errorMessage = {
                error: "Error sending screenshot to
                servlet."
            };
```

```
                    // Encapsulate the error message in a JSON-
                    formatted String and send it over the WebSocket
                    connection.
                    connection.send(JSON.stringify(errorMessage));
                });
            }
        });
                // Set the number of received chunks to zero, as a reset
                for the next screenshot.
                receivedChunks = 0;
            }
        }
    });

    // Use the "close" keyword to handle client disconnection.
    connection.on("close", function (reasonCode, description) {
        // Log a message to indicate that the client has disconnected.
        console.log("Client has disconnected.");
         // Stop the monitoring process when the client disconnects.
        clearInterval(monitoringInterval);
    });
});
```

Adding the NTMA Block

The code for the NTMA process is packed in the monitorNetworkTraffic() method,
which is called when the two input parameters from the client side are successfully
received. This method is given in Listing 7-17.

Listing 7-17. Adding the NTMA block

```
// Create a method for the NTMA process.
function monitorNetworkTraffic() {
    // Initialize a counter to know when enough network traffic has been
    collected.
    var counter = 0;
```

```javascript
// Initialize an array to contain latency items.
var latencyArray = [];

// Initialize a counter to know when enough latency values have been
collected.
var counterLatency = 0;

// Create a function to be run every second using the
setInterval method.
// This method takes as input a function and the time interval in ms.
monitoringInterval = setInterval(function() {
    // Use the inetLatency() built-in method to read latency values.
    si.inetLatency().then(data => {
        // Check if latency cannot be read, which would mean that there
        is no internet connection.
        if (data == null) {
            // Create an error message to be sent over to the client.
            const errorMessage = {
                error: "No Connection"
            };
            // Encapsulate the error message in a JSON-formatted String
            and send it over the WebSocket connection.
            connection.send(JSON.stringify(errorMessage));
            // Stop the monitoring process.
            clearInterval(monitoringInterval);
        // If connection is OK.
        } else {
             // Fill the latencyArray object with latency values using
             the counterLatency counter.
            latencyArray[counterLatency] = data;
            // Specify that the latency equals the data read from the
            Wi-Fi interfaces.
            latency = data;
        }
        // When a total of 60 values is recorded, one minute of traffic
        has been collected.
```

```
    if (counterLatency == 59) {
        // Call the method to calculate the QoS based on the
        latency values recorded.
        scoreQoS = calculateQOS(latencyArray);
    }
    // Increment the counterLatency object.
    counterLatency++;
})

// Fetch network data from Wi-Fi interfaces.
si.networkStats().then(data => {
// Calculate the number of packets received per second, equivalent
to the download speed.
var RX = Math.round(((((data[0].rx_sec)/1000) + Number.EPSILON)
* 100) / 100;
// Calculate the number of packets sent per second, equivalent to
the upload speed.
var TX = Math.round(((((data[0].tx_sec)/1000) + Number.EPSILON)
* 100) / 100;

// Add newly read traffic to pre-loaded arrays using the
push() method.
// The pre-loaded arrays contain parameters stored in a time-series
fashion.
// This may increase the performance of the algorithm depending on
the device activity.
downloadArray.push(RX); // Add the download speed value to its
                           pre-loaded array.
uploadArray.push(TX); // Add the upload speed value to its
                         pre-loaded array.

// Log the values into console.
console.log("Download Speed: " + RX + " KB/s");
console.log("Upload Speed: " + TX + " KB/s");
console.log("-------------        --------");

// When a total of 60 values has been recorded, the monitoring
process is stopped and the ML algorithms are triggered.
```

437

```
if (counter == 60) {
    // Stop the monitoring process using the
    clearInterval() method.
    clearInterval(monitoringInterval);

    // Call the prediction method for making time-series prediction
    for download speed.
    // Ensure that the method is called again in case the algorithm
    fails, and returns a NaN.
    while (true) {
        predRX = predictMLP(downloadArray, parseInt(sampleSize),
        parseInt(predTime), 0.1, 0.01);
        if (!isNaN(predRX)) {
            // Break out of the loop if an actual number is
            obtained.
            break;
        }
    }
}

    // Call the prediction method for making time-series prediction
    for upload speed.
    // Ensure that the method is called again in case the algorithm
    fails and returns a NaN.
    while (true) {
        predTX = predictMLP(uploadArray, parseInt(sampleSize),
        parseInt(predTime), 0.1, 0.01);
        if (!isNaN(predTX)) {
            // Break out of the loop if an actual number is
            obtained.
            break;
        }
    }

    // Call the prediction method for making time-series prediction
    for latency.
    // Ensure that the method is called again in case the algorithm
    fails and returns a NaN.
```

```javascript
    while (true) {
        predLat = predictMLP(latencyArray, parseInt(sampleSize),
        parseInt(predTime), 0.1, 0.01);
        if (!isNaN(predLat)) {
            // Break out of the loop if an actual number is obtained.
            break;
        }
    }

    // Classify the device activity by calling the
    // classification method.
    deviceActivity = calculateKNN(RX, TX);

    // Create a constant variable holding the prediction and
    // classification results obtained.
    const predictionMessage = {
        downloadSpeed: RX,
        uploadSpeed: TX,
        predictedDownloadSpeed: predRX,
        predictedUploadSpeed: predTX,
        predictedLatency: predLat,
        predictedQoS: scoreQoS,
        deviceActivity: deviceActivity,
        averageLatency: averageLatency
    };
    // Encapsulate the metrics message in a JSON-formatted String
    // and send it over the WebSocket connection.
    connection.send(JSON.stringify(predictionMessage));
}
// If the number of values collected has not yet reached 60.
else {
    // Create a constant variable containing the network traffic
    // values to be sent to the client every second.
    // Add all 3 real-time values.
    const jsonMessage = {
        downloadSpeed: RX,
```

```
                uploadSpeed: TX,
                latency: latency
            };

            // Encapsulate the real-time message in a JSON-formatted String
            and send it over the WebSocket connection.
            connection.send(JSON.stringify(jsonMessage));
        }
        // Increment the network traffic counter.
        counter = counter + 1;
        })
    }, 1000) // Specify an interval of 1000 ms for the monitoring process.
}
```

7.3.8 Prediction, Classification, and Network Score Computation Methods

With VQA completed, the prediction, classification, and network score computation methods are performed by the `predictMLP()`, `calculateKNN()`, and `calculateQOS()` methods, respectively, which contain the exact same codes as given in Sections 5.3.7, 5.3.8, and 5.3.9. They are thus added to the `server.js` file.

7.4 Integrated NTMA/VQA Application Testing and Deployment

This marks the completion of the integrated NTMA and VQA application where the same servlet used for the VQA application can be hosted with the paths to the screenshot folders changed accordingly. The client, server, and servlet frameworks are now ready to be tested in varying network and video streaming environments. The server and servlet are started using the same principles explained in Section 6.5. The integrated application is thus tested in different network scenarios in a typical home setting, as shown in Figures 7-8 to 7-13.

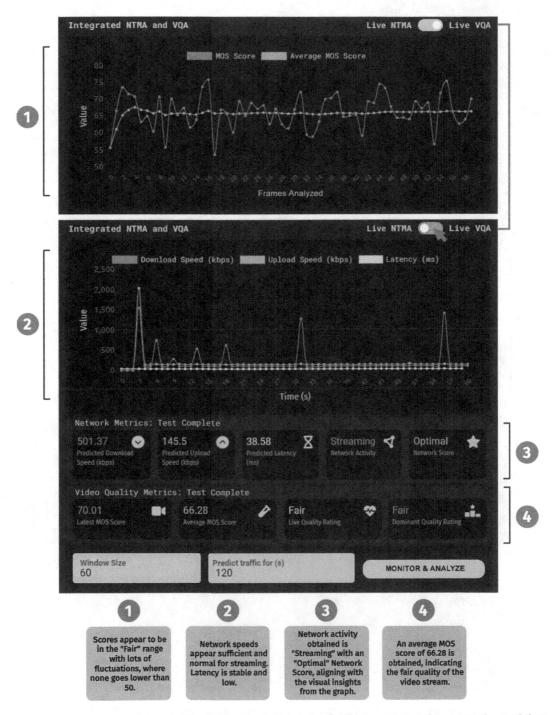

Figure 7-8. *Streaming on YouTube in 480p, resulting in optimal network and fair video quality*

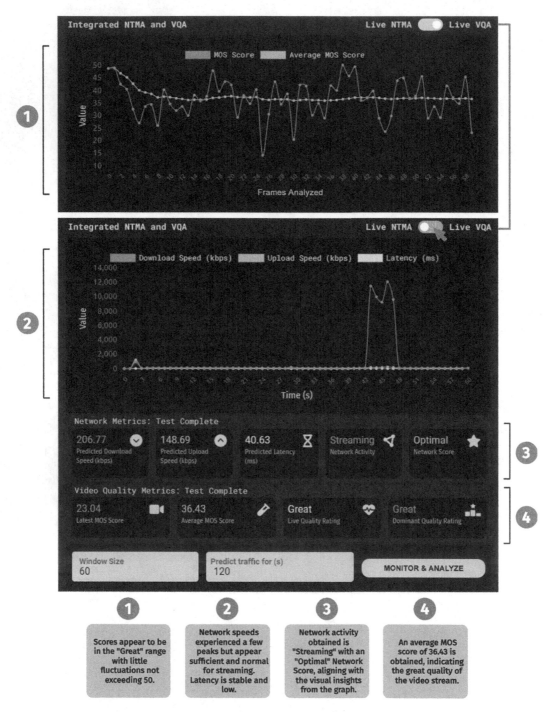

Figure 7-9. *Streaming on YouTube in 1440p, resulting in optimal network and great video quality*

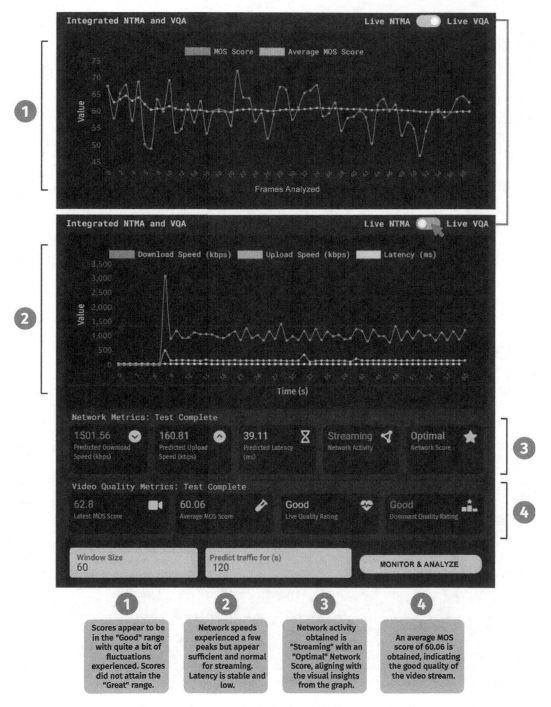

Figure 7-10. *Streaming on Twitch in 720p, resulting in optimal network and good video quality*

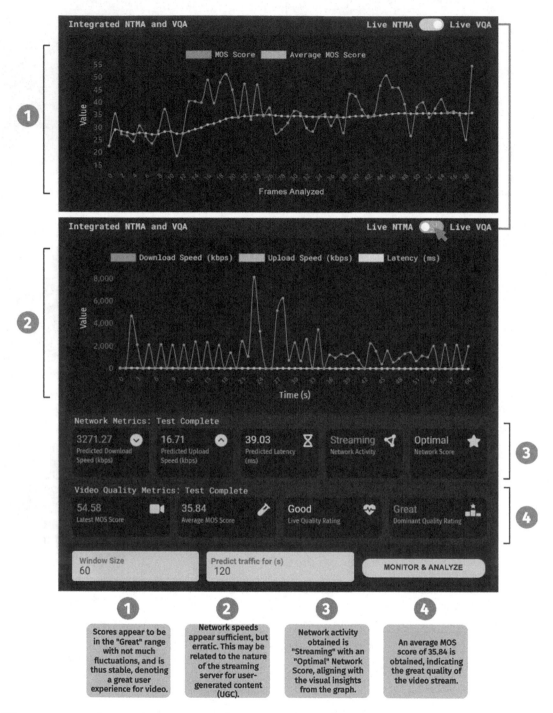

Figure 7-11. *Streaming on Twitch in 1080p, resulting in optimal network and great video quality*

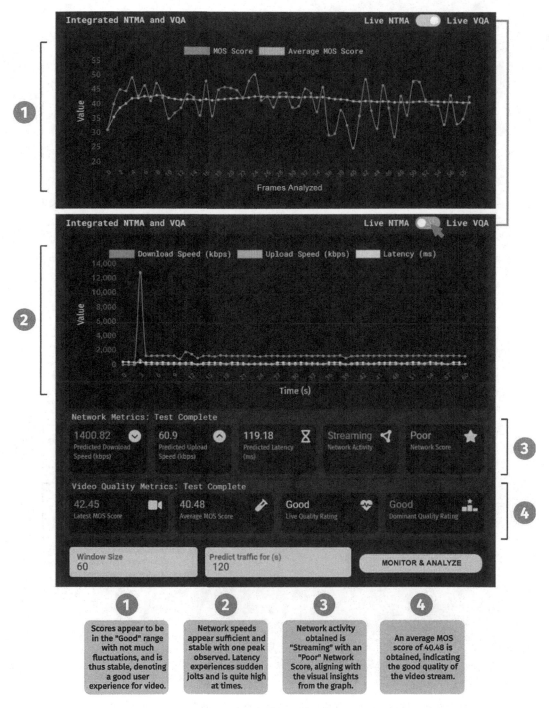

Figure 7-12. *Streaming on Twitch in 1080p, resulting in poor network and good video quality*

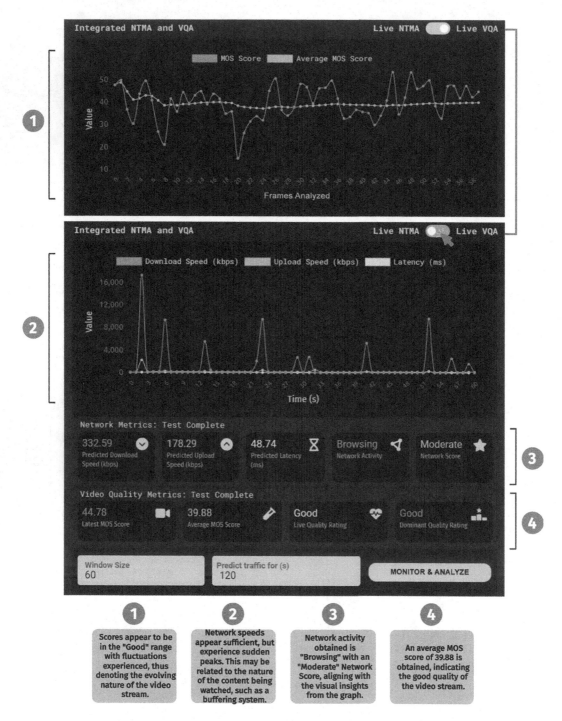

Figure 7-13. *Streaming on TikTok, resulting in moderate network and good video quality*

7.5 Summary

With both the NTMA and VQA applications completed, Chapter 7 delved into their integration, allowing for both functionalities to be available in a single application inside the web browser. Therefore, a combination of the files used as the client, server, and servlet for the two independent processes are properly modified to accommodate a changed layout and a button that performs both functions. This chapter made the connection between the two spheres, allowing the end user to obtain a top view of network performance in real time, with predictive analytics, device activity classification, a network QoS score, and a video quality score. Some key takeaways are given next:

- A unified system may be used to monitor and analyze network traffic and video quality in a comprehensive manner, with real-time capabilities.

- The integrated NTMA and VQA application consists of a system model, has client, server, and servlet paradigms, and follows the same principles described in previous chapters.

- As a future work, additional techniques can be employed for comparison and evaluation against the existing algorithms. For example, other ML algorithms like support vector machines (SVM), random forests, or gradient boosting machines (GBM) could be explored to assess their performance in NTMA and VQA tasks.

- The integrated NTMA and VQA application can be used to analyze different types of networks, such as 4G (LTE), 5G, and Wi-Fi networks, where the metrics serve as statistics for further analysis. With network traffic parameters anticipated in real time, sudden dips in the download and upload speeds can be investigated by telecommunication operators before the issue actually occurs.

- The deployment of the integrated application throughout several PCs in a network environment results in the possibility of carrying out multiple tests to gauge its performance by observing the network and video scores.

- The application can serve as a first-layer diagnostic tool to identify deviations from the norms in metrics such as jitter and the MOS score, which can be used by telecommunications companies to immediately locate which metric to act upon, investigate, and prioritize.

- Regulatory bodies can use the integrated application as a means to perform unbiased QoS measurement. – Chapter 7

7.6 References–Chapter 7

1. Google, "Google Fonts," Google Fonts, 2019, `https://fonts.google.com/` (accessed Dec. 23, 2023).
2. Font Awesome 5, "Font Awesome 5," Fontawesome.com, 2017, `https://fontawesome.com/` (accessed Dec. 23, 2023).
3. M. Otto, "Bootstrap," Bootstrap, 2022, `https://getbootstrap.com/` (accessed Dec. 23, 2023).
4. chart.js, "Chart.js | Open source HTML5 Charts for your website," Chartjs.org, 2019, `https://www.chartjs.org/` (accessed Dec. 23, 2023).
5. "bootstrap," cdnjs, `https://cdnjs.com/libraries/bootstrap/5.0.0` (accessed Mar. 26, 2024).
6. "Chart.js," cdnjs, `https://cdnjs.com/libraries/Chart.js/3.8.0` (accessed Mar. 26, 2024).
7. B. McKelvey, "websocket," npm, Apr. 14, 2021, `https://www.npmjs.com/package/websocket` (accessed Dec. 23, 2023).
8. Node.js, "HTTP | Node.js v18.8.0 Documentation," nodejs.org, `https://nodejs.org/api/http.html` (accessed Dec. 23, 2023).
9. S. Hildebrandt, "systeminformation," npm, Dec. 22, 2023, `https://www.npmjs.com/package/systeminformation` (accessed Dec. 23, 2023).
10. Node.js, "File system | Node.js v18.1.0 Documentation," nodejs.org, `https://nodejs.org/api/fs.html` (accessed Dec. 23, 2023).
11. U. Biallas, "mlp," npm, Jun. 30, 2016, `https://www.npmjs.com/package/mlp` (accessed Dec. 23, 2023).
12. ml.js, "ml-knn," npm, Jun. 29, 2019, `https://www.npmjs.com/package/ml-knn` (accessed Dec. 23, 2023).
13. J. Richardson, "fs-extra," npm, Nov. 28, 2023, `https://www.npmjs.com/package/fs-extra` (accessed Feb. 11, 2024).

14. L. Fuller, "sharp," npm, Jan. 2024, `https://www.npmjs.com/package/sharp` (accessed Feb. 12, 2024).

15. Axios, "axios," npm, Jan. 25, 2024, `https://www.npmjs.com/package//axios` (accessed Feb. 11, 2024).

16. "Reddit," `https://www.reddit.com/` (accessed Feb. 18, 2024).

17. "Instagram," `https://www.instagram.com` (accessed Feb. 18, 2024).

18. "Twitch," `https://www.twitch.tv` (accessed Feb. 18, 2024).

19. "YouTube," `https://www.youtube.com` (accessed Feb. 18, 2024).

Index

A

Absolute Category Rating (ACR)
approach, 20, 109
AccepTV video quality monitor, 118
Access control lists (ACLs), 72
Active monitoring, 9
Adaptive bit rate (ABR), 128–129, 255
Adaptive streaming technologies, 26
addData() function, 287
addData() method, 208, 213, 282, 287, 419
Anomaly detection, 5, 10, 34
Apache Tomcat Java servlet, 379
Architectural framework, 58
Artificial intelligence (AI), 27
Artificial neural networks (ANNs), 34
atof()and atoi() methods, 363
Attention and focus, 102
Automated multimedia traffic
management framework, 82
Autonomous decision-making, 67
Autoregression moving
average (ARMA), 84, 86
Autoregressive integrated moving
average (ARIMA), 2, 86
Autoregressive moving
average (ARMA), 84, 86
axios library, 299

B

Background script functions, 185, 260

Benchmarking methodologies
objective metrics as reference
points, 111
standardized testing protocols, 111
BERT, 35, 36
Binarization process, 159, 165
Bitstream-based assessment, 21
Blind/Reference-less Image Spatial
Quality Evaluator (BRISQUE)
method, 157, 158, 172
Block-based techniques, 19
Blurring, 19

C

calculateKNN method, 215
calculateQoS() method, 215, 224, 236
Centralized cloud, 67
Chart.js, 203
Chrome extension API, 201
Cisco's findings, 78
Citrix Systems, 69
Classification model, NTMA
browsing, 142
data collection, 142, 143
data preparation, 144
device activity, 142
idle, 142
KNN, 144
sample data, KNN
classification, 145–146
streaming, 142

© Tulsi Pawan Fowdur, Lavesh Babooram 2024
T. P. Fowdur, L. Babooram, *Machine Learning For Network Traffic and Video Quality Analysis*,
https://doi.org/10.1007/979-8-8688-0354-3

W, X

Printed in the United States
by Baker & Taylor Publisher Services

Printed in the United States
by Baker & Taylor Publisher Services